WOLFRAM
SUMMER SCHOOL
RESEARCH REPORTS
2023

Edited by Mohammad Bahrami

Wolfram Summer School Research Reports 2023

Edited by Mohammad Bahrami
Copyright © 2025 by Wolfram Media, Inc.

Wolfram Media, Inc.
wolfram-media.com
ISBN 978-1-944183-98-1 (paperback)
ISBN 978-1-944183-99-8 (ebook)

Typeset with Wolfram Notebooks: wolfram.com/notebooks

First edition.

Table of Contents

Preface

MOHAMMAD BAHRAMI
Director of Academic Innovation Support

The Wolfram Summer School is a unique experience for students and mentors alike. In these proceedings, you will see some of the fruits of the 2023 Summer School. Before spotlighting the projects, it is worth highlighting the collaborative environment and culture that is the "secret sauce" of the school.

The first ingredient is the diversity of backgrounds that come together for the Summer School. Geographic diversity is always a given, with the 2023 Summer School hosting students from over 28 countries. Students at the school also range in experience from undergraduates to seasoned faculty and industry professionals. There are always a wide range of academic backgrounds, knowledge and skills represented.

Mentors at the school come from a similarly wide variety of backgrounds. In fact, it is not an uncommon story for mentors to have been students in the past. This range of experiences represented in the students and the mentors allows for support in both the mentor-student interactions and in peer assistance between students. The network of personal connections formed during the Summer School is an exciting source of future collaborations.

The next ingredient for the success of the school is the way mentorship is structured. Mentors are most often veteran Wolfram developers, but can also include longtime friends of the school. Given the interests of each student, a small group of two to three mentors assumes the responsibility of being that student's primary source of guidance through the school. However, the school very much follows the philosophy that "it takes a village" to achieve the best possible projects.

While each student works most closely with "their" mentors, the school provides a collaborative space that can quickly pull in experts from all over Wolfram Research to address particular needs. From casual interactions at meals, to structured presentations, to project meetings led by Stephen Wolfram, to late-night discussions on philosophy and strategy, there is no better way for students to see "how things get done" at Wolfram Research.

The last ingredient for success is the project-based approach. The goal of the school is for each student to go beyond a simple certificate on their resume and create something *real* that can be immediately useful to them, a demonstration of the skills they have acquired and a starting point for their future plans.

The collection of projects here is necessarily only a fraction of the interesting directions taken by students. However, the projects do illustrate some common themes that emerged:

- By 2023, large language models (LLMs) were prominent in the news and everyone's mind, and so a number of projects in the Science & Technology track explored these tools. Some projects addressed the science of LLMs, aiming to understand the patterns that emerge at various layers of the models. Others were directed at applications of LLMs, including for use cases in the education sector.

- New work continued on the Wolfram Physics Project, with projects exploring new features, phenomenology and ways to understand this class of models. A number of related projects studied multiway systems in contexts outside of physics.

- Continuing the spirit of *A New Kind of Science*, many projects engaged in ruliology, or the study of the complex evolution of systems that proceed according to simple rules. Each school brings the opportunity for discovering new mathematical (and metamathematical) results.

- In every iteration of the Wolfram Summer School, a number of projects seek to expand both convenience and core functionality for the community of Wolfram Language users. Several of the projects selected here implement new algorithms, functions and packages with an eye for the end user.

In sum, the Wolfram Summer School is a unique blend of intellectual curiosity, scientific research, professional development and entrepreneurial insights. The projects selected here are the result of a collaborative process that often leads to fruits beyond the celebratory poster sessions and graduation ceremony.

PROJECT LIST

The Project List is a catalog of abstracts for all student projects from the Wolfram Summer School 2023. Scan the QR code or visit the URL on each page to view the full project and interact with the code.

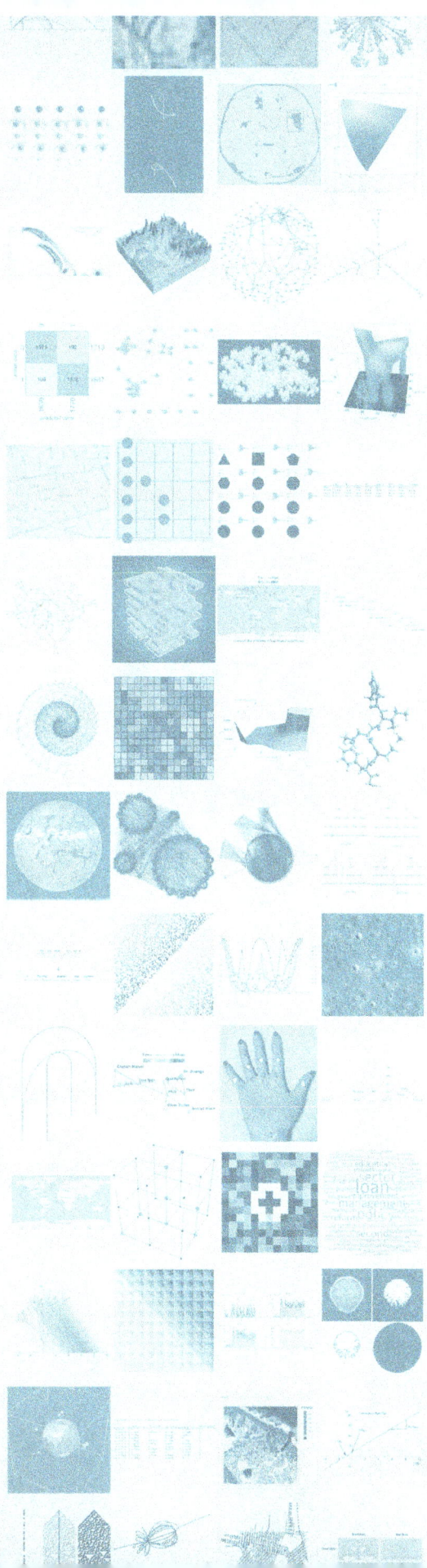

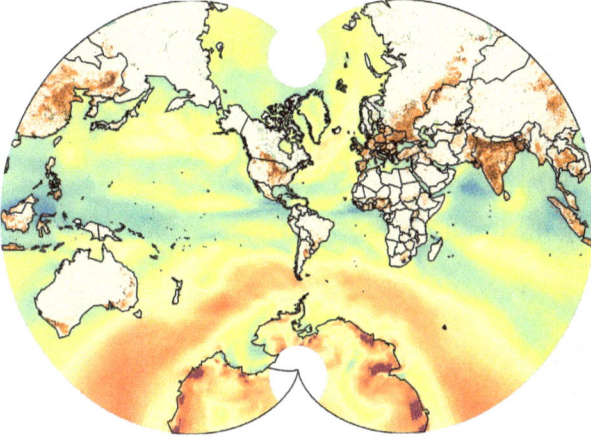

RemoteSensing Paclet: GIBS Satellite Imagery and AppEEARS Geographic Data Products

PHILEAS DAZELEY-GAIST

This project aims to develop robust tools for the integration of GIBS satellite imagery and AppEEARS geographic data products into Wolfram Language. The focus of this project is to build efficient and flexible methods to import, visualize and interpret this geographic data using Wolfram Language as well as examples and documentation detailing the methods involved. We are delighted to announce the release of the RemoteSensing paclet to the Wolfram Language Paclet Repository as the culmination of these three weeks of work at the Wolfram Summer School. We will continue to work on this repository after the Summer School, and look forward to your feedback!

Scan or visit
wolfr.am/WSS2023-Dazeley-Gaist

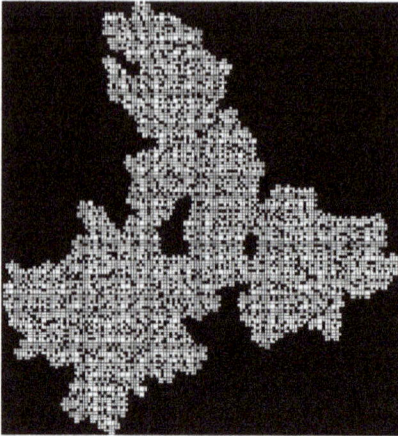

Efficient Discovery of Halting Paths in Aggregation System Multiway Graphs

PIETRO PEPE

This project investigates totalistic aggregation systems, which exhibit cellular automata behavior with a unique twist. The focus is on totalistic rules that determine cell eligibility based on the total number of active neighboring cells. By developing a simulation software using Lua and the LÖVE framework, the project enables the exploration of different rule sets and initial conditions. Multiway graphs are used to visualize the system's behavior, and the concepts of translation, rotation and reflection canonicalization simplify the analysis of equivalent states. The project also proposes a new classification system—k-bit rules—that systematically identifies halting states based on rules and minimal initial conditions, and uncovers insights into system growth and eventual halting. Overall, this project provides valuable insights into totalistic aggregation systems and their dynamics.

Scan or visit
wolfr.am/WSS2023-Pepe

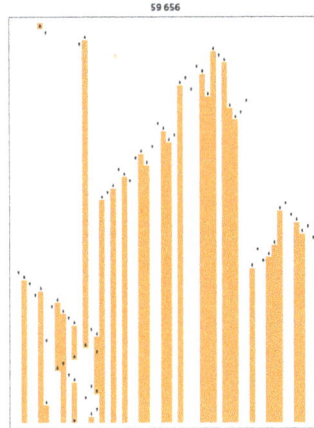

Turing Machines on Graphs

STEFAN GRAHAM

Ordinary Turing machines are implemented on infinite 1D tapes. This project investigates and explores the behavior of Turing machines on finite cyclic graphs. The graphs investigated are the circular cyclic graph, a torus and the Sierpiński network. This post demonstrates that the Sierpiński network exhibits the most complex and varied behavior.

Scan or visit
wolfr.am/WSS2023-Graham

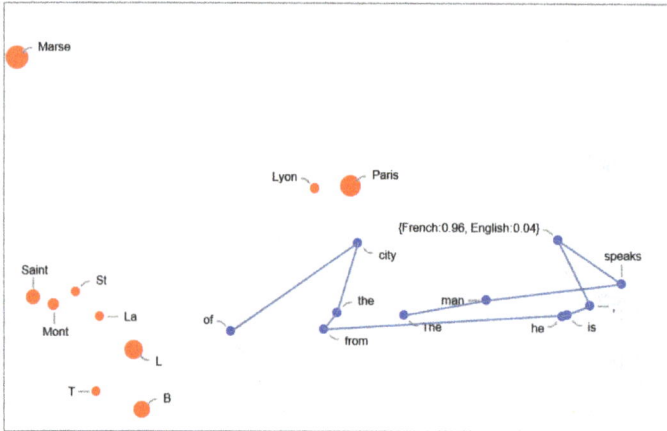

Study the Effect of Small Changes in Large Language Models

LUHAN CHENG

Large language models (LLMs) have demonstrated their strong capabilities in language modeling, text generation and many other natural language understanding tasks compared to their predecessors. We hypothesize LLMs gain their strength by successfully mapping words onto the semantic space and traversing through the semantic space with a high degree of stability. We model the computation of LLMs as trajectories in their embedding spaces. We study the stability of the trajectory by introducing small amounts of controlled noise into the input word embeddings. We demonstrate that GPT-3 is more capable of resisting small perturbations in the input space compared to the GPT-2 model.

Scan or visit
wolfr.am/WSS2023-Luhan

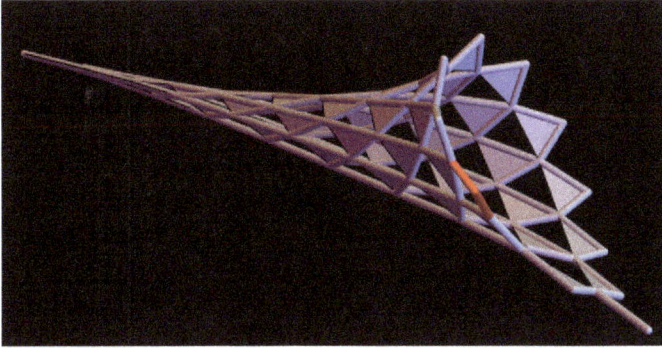

Animating Wolfram Model Evolutions in 3D

DUGAN HAMMOCK

The evolution of hypergraphs undergoing successive rewriting events is a central theme for the Wolfram Physics Project. Individual hypergraphs are typically plotted using the GraphLayout *option* "SpringElectricalEmbedding", *which preserves and displays the overall shape and topology of the hypergraph. Animations of Wolfram models undergoing successive updating events have historically suffered from instability in the placement of vertices from one frame to the next. In this project, an algorithm is implemented that solves this problem and computes vertex coordinates for hypergraph animations in a temporally coherent manner.*

Scan or visit
wolfr.am/WSS2023-Hammock

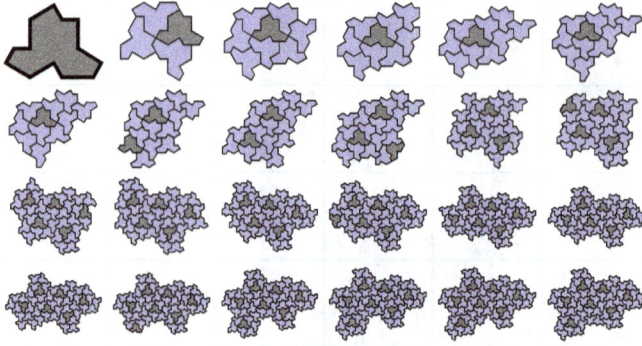

Hat Tiling Space Reduction and Grow Function Implementation

BOWEN PING

This project aims to implement a function to grow a given hat tile cluster automatically. At first, some reduction about specific vertex configurations are deducted to filter possible cases. Then a multiway grow function is implemented to study some special hat tile clusters. A dynamic module showing the grow process of clustering is implemented to illustrate the procedure with clarity. Additionally, a function with a fast algorithm to find all special subclusters in a big super-tile is obtained. The next step is to use more local vertex information to improve the growth algorithm, which will be useful in studying the hat tile space.

Scan or visit
wolfr.am/WSS2023-Ping

Indicates Correct Score

LLM-Powered Reviews of Student Work Samples

AARON CARVER

This project uses LLMs to review student work samples, specifically ACT writing essays. First, I explore using LLMs to apply a descriptive grading rubric that maps descriptions of writing quality to quantitative scores. Second, I use LLMs and the approach of few-shot learning to perform grammatical error correction (GEC), which is the process of identifying and fixing errors in the categories of spelling, punctuation, grammar and word choice.

For most of the writing samples, the LLM was able to accurately assess the correct score; however, each test produced some inaccurate and inconsistent results. I see several opportunities for further experimentation, which may lead to performance adequate for practical use such as varying temperature, preening effective examples for few-shot learning and implementing multiple calls to LLMs to disambiguate scores that vary by a single point.

For the GEC task utilizing few-shot learning, the LLMs showed a generally powerful ability to find errors and return them in a structured data format. However, the LLMs also "hallucinated" several false errors. In practice, a combination of traditional GEC tools with LLM-powered reviews will likely lead to the best results.

LLMs currently have the power and flexibility to provide useful reviews of student work samples, but the tools that I believe will succeed in practice will be those that add additional computational power and human judgement to LLM-powered workflows.

Scan or visit
wolfr.am/WSS2023-Carver

From Elliptic Curves to Diophantine Equations: A Journey through Rational Points

ADITI KULKARNI

Elliptic curves, commonly represented as $y^2 = x^3 + Ax + B$ (where A and B are constants) involve interactions among geometry, number theory and algebra. This project explores rational points on elliptic curves. By repeatedly applying a geometric procedure (the tangent-secant method) to a finite set of solutions, all of the possibly infinitely many rational solutions to an elliptic curve equation may be found. We will examine the relationship between integer solutions of Diophantine equations and rational points on elliptic curves. The simple-looking Diophantine equation $\frac{a}{b+c} + \frac{b}{a+c} + \frac{c}{a+b} = n$ (where n is a fixed positive integer) is infamous for having very large minimal solutions. This homogenous equation in three variables can be transformed to an elliptic curve in a two-dimensional plane. Additionally, rational points on the elliptic curve lead us to least positive integer solutions of the Diophantine equation. We will be learning about solutions of $\frac{a}{b+c} + \frac{b}{a+c} + \frac{c}{a+b} = 6$ in this project.

Scan or visit
wolfr.am/WSS2023-Kulkarni

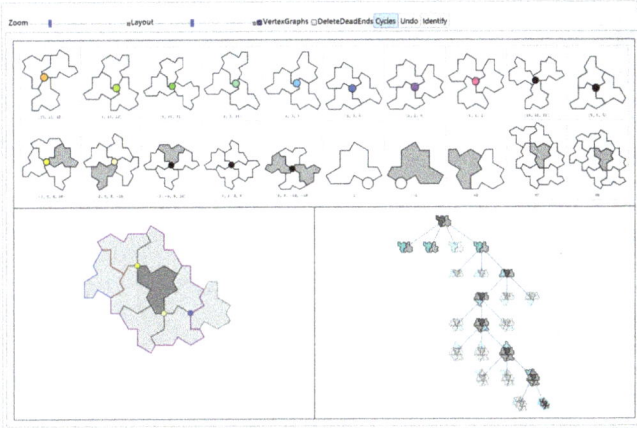

HatGame: A Journey through the Hat Tile Configuration Space

JOHANNES MARTIN

A central objective of this project is the creation of an interactive (one-player) game to explore properties of the hat tile. One can combine jigsaw-like hat tiles into larger and larger clusters. The game contains a few tools that help to analyze the structure of such clusters. First explorations with the HatGame were helpful to cut down the number of possible initial configurations and allowed for extension of "the ten-vertex theorem" of [1]. The hat tile is the first solution that has been found for the so-called "einstein problem," which was the search for one single shape (tile) that can fill the entire plane with only aperiodic patterns (tiling). A tiling is valid and aperiodic when it has the following properties:

1) It covers the entire plane without gaps or holes.
2) There is no translation that maps the pattern onto itself, i.e. there are no translational symmetries.

Remark: a tile is an "Einstein" (an aperiodic mono-tile) when ALL its valid tilings are aperiodic.

Scan or visit
wolfr.am/WSS2023-Martin

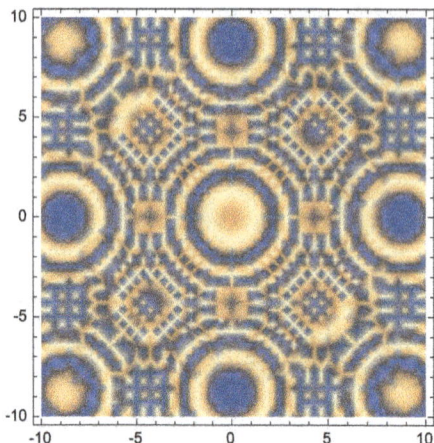

Blackbox Optimization: The Mesh Adaptive Direct Search (MADS) Algorithm

MOISES GONZALEZ

Blackbox optimization is a branch of optimization that deals with derivative-free functions. This means that the function's derivative is unknown because finding it is impossible or impractical. Different methods—like random search, grid search and some heuristics—are used to solve these optimization problems. One recent approach used in this area is the mesh adaptive direct search (MADS) method. This paper aims to explain the creation of a function in Mathematica that runs the MADS algorithm and returns a minimum. This function has been successfully implemented, throwing good results in global minima searches. Nevertheless, using it in functions with many local minima is not suggested because it might get trapped in one of these minima. Also, in the examples, there are functions with problematic derivatives that are successfully solved. The function was named MadsOptimizer, and it takes two parameters. The first parameter is the objective function, and the second is a list of the variables. Given the positive results of the MADS implementation, the code can be turned into a built-in Wolfram Language function that can be used in different areas. MADS is widely used for parameter fit and other problems involving simulations and experiments. Also, this project suggests an addition to the MadsOptimizer function, increasing the chances of getting the global minima. This paper grasps some of the theory behind the method, shows some results using different functions and uses a visual representation to show the path taken by the algorithm.

Scan or visit
wolfr.am/WSS2023-Gonzalez

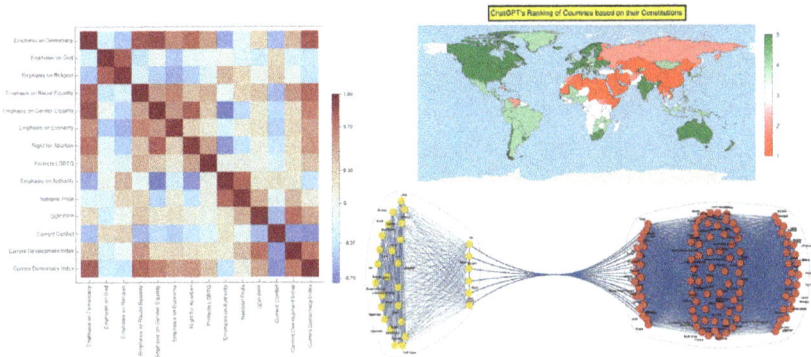

Using Large Language Models to Generate Insights from World Constitutions

JOEL ABDULAI KALLON

This project leverages Wolfram's large language model (LLM) capabilities to explore how a country's constitution influences its socioeconomic and development outcomes. Using OpenAI's GPT-3.5 in a few-shot learning approach, countries were classified into governance types and assessed on the emphasis of values like democracy, religion, economic development, equality and state authority.

Key findings include that more religious countries tend to be poorer, and countries emphasizing democracy tend to rank higher in the Global Democracy Index, demonstrating GPT-3.5's effective classification capabilities.

In conclusion, this study showcases the potential of LLMs for deriving insights from extensive text data. GPT-3.5 provided a competent assessment of constitutional values across 210 countries, revealing correlations such as the negative relationship between religious emphasis in constitutions and GDP per capita.

Scan or visit
wolfr.am/WSS2023-Kallon

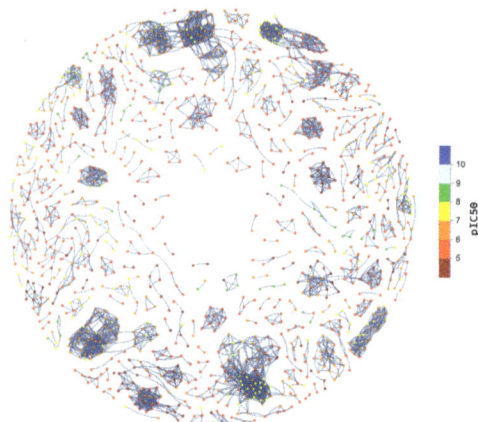

Chemical Space Networks from Molecule Data

JOSHUA KHORSANDI

This project develops two interrelated tools for the generation and exploration of chemical space networks from molecule data; a Wolfram Language superfunction ChemicalSpaceNetwork that generates chemical space graphs from imported data and includes options for a wide range of distance functions and molecular fingerprints while integrating with the MolecularFingerprints package; and a graphical user interface designed to be used with the superfunction that easily enables the analysis of complex chemical network data. Future work in this area includes updates to the interface as well as applications in random matrix theory and graph neural networks to areas of medicinal chemistry, cheminformatics and drug discovery.

Scan or visit
wolfr.am/WSS2023-Khorsandi

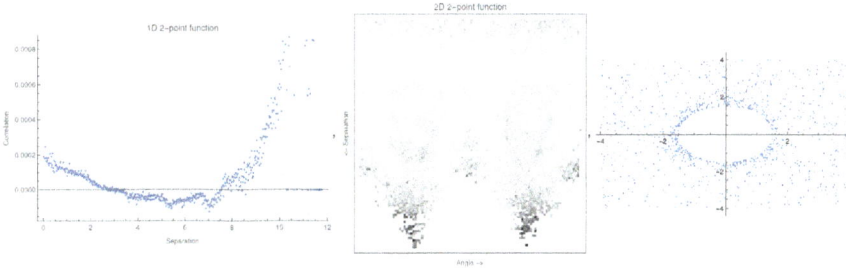

Cosmological Imprints on the Two-Point Correlation Function of Over-Dimension Fields

FEDERICO SEMENZATO

The aim of this project is to investigate the statistical features of over-dimension fields in a cosmological setting. The over-dimension field of a graph is defined as the excess Hausdorff dimension computed at each vertex with respect to the global average. The two-point correlation function has become one of the most popular statistical tools in cosmology. We build an estimator in a graph-based framework and apply it to over-dimension fields. First, this is done by considering graphs generated from sprinkling of known geometries. The impact of using coordinates or geodesics to compute vertex separation is addressed and the formalism is then extended to higher-order statistics (relevant for non-Gaussian fields). We apply this approach to Gravitas generated hypergraphs with an underlying Friedmann–Lemaître–Robertson–Walker (FLRW) metric in which some local dimension perturbation is introduced, and the main resulting features are discussed. This is relevant to the construction of observational predictions in the context of the Wolfram Physics Project, and potentially to the prediction of signatures in the large-scale structure of the universe and on the cosmological stochastic gravitational-wave background anisotropies.

Scan or visit
wolfr.am/WSS2023-Semenzato

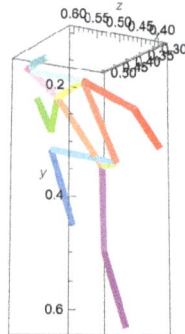

3D Human Pose Estimation Using Machine Learning

FIZZA RUBAB

Human pose estimation is the task of estimating the positions and orientations of a person's body joints from images or videos. This project attempts to tackle the three-dimensional variant of this problem using state-of-the-art computer vision techniques. The CenterNet model is used to obtain 2D keypoints from image frames, estimate the depth of each keypoint using the MiDaS depth estimation model imported via ONNX and, finally, scale and visualize the 3D skeletons in an interactive display.

Scan or visit
wolfr.am/WSS2023-Rubab

Looking at Network Packet Data with Wolfram Language and GPT Models

JOHN ADAMS

"You are an epic poet who is telling the story of network packet data in a poem in the style of the hero's journey. Generate network packet interface data and return it in CSV format. Include values of the fields of simulated data in the poem."

For our investigation, we will use Wolfram Language LLM functions with various prompts to examine network packet data. We will also use ImageSynthesize to generate images based upon simulated network packet data generated using LLMSynthesize. These functions will require importing the LLM paclets for Wolfram Language. We will also need an OpenAI API key for the LLM functions. See the following links:

- reference.wolfram.com/language/ref/NetworkPacketCapture.html

- resources.wolframcloud.com/PacletRepository/resources/Wolfram/LLMFunctions

- platform.openai.com/docs/api-reference

Scan or visit
wolfr.am/WSS2023-Adams

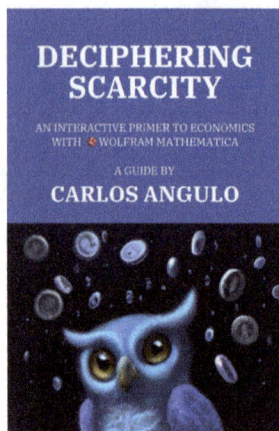

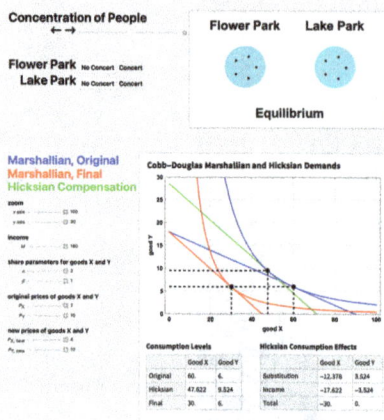

Shaping the Future of Economics Education: An Undergrad's Venture into Interactive Pedagogy

CARLOS MERARDO ANGULO ZUMAETA

Conventional teaching and learning methodologies utilising non-interactive material can fall short of engaging students and fostering intellectual curiosity. This project presents an interactive ebook designed to reshape the pedagogical landscape of economics by showcasing vivid storytelling, insightful example problems from real-life situations and Wolfram Mathematica interactive animations. Strategically structured into seven key sections, the ebook guides the reader from defining key principles to undergraduate-level practice material, prioritising the careful explanation of core economic intuition rather than complex terminology. Leveraging interactive elements coded using the Manipulate *function, the book encourages problem solving and critical thinking, leading to a deeper understanding of economics through an engaging learning journey. Future work includes addressing areas for improvement, completing the ebook, planning and execution of distribution avenues and marketing campaigns.*

Scan or visit
wolfr.am/WSS2023-Zumaeta

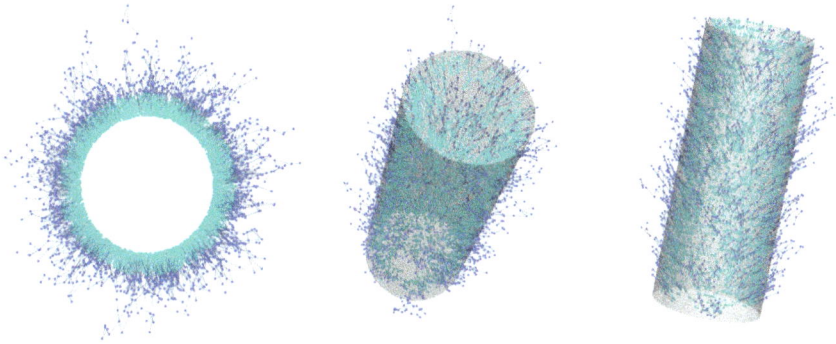

Black Hole Entanglement Entropy from Causal Graphs

JACOPO UGGERI

Entanglement entropy on causal graphs is suspected to be related to entropy calculations on Wolfram model branchial graphs, the latter being more computationally expensive. The project aims to compute the entropy of a Schwarzschild black hole leveraging the established formalism of causal set theory for calculating entanglement entropy on causal graphs, with the ultimate goal of testing whether the area scaling law holds for discrete spacetime models. Black hole causal graphs were obtained by discretizing spacelike hypersurfaces and computing the causal relations among events in consecutive hypersurfaces.

Scan or visit
wolfr.am/WSS2023-Uggeri

Exploring Random Quantum Circuits without Unitary Matrix Decomposition

CHARLES WOODRUM

In this notebook, we explore the properties of random quantum circuits. Random quantum circuits can be derived in a variety of ways. The most straightforward way is to create a unitary matrix of a desired dimension (in this case, 2^n, where n is the number of qubits in the system), then turn that matrix into an operator on the circuit. However, this process is not straightforward to implement on a quantum device, since the unitary matrix must be broken down into operations that can be implemented on quantum hardware.

Scan or visit
wolfr.am/WSS2023-Woodrum

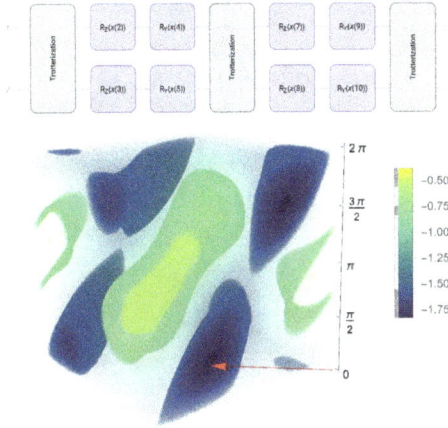

Developing a Variational Quantum Eigensolver

ARIELLE SANFORD

What Is a Variational Quantum Eigensolver?

Variational quantum eigensolvers (VQEs) use both classical and quantum computing to solve for the ground state of a given Hamiltonian, an operator that gives the energy of a system. Once the Hamiltonian is ascertained, one must determine an appropriate ansatz. A VQE ansatz is typically composed of a series of parametrized gates on a quantum circuit. These gates are initialized to the zero state, but with cumulative modification to the parameters will ideally give the ground state of the Hamiltonian. Next, all VQEs use some sort of classical minimization routine. Here I will use Mathematica's NMinimize, *which automatically determines an optimal routine for the problem and numerically searches for the global minimum. The quantum computing component to VQE is the cost function of this routine. The cost function takes in a list of parameters and the desired ansatz, computes the quantum state and outputs the expectation value of this state with respect to the Hamiltonian. With enough iterations, a good minimization routine and a good ansatz, we should be able to find the exact ground state of the system along with its corresponding energy.*

Scan or visit
wolfr.am/WSS2023-Sanford

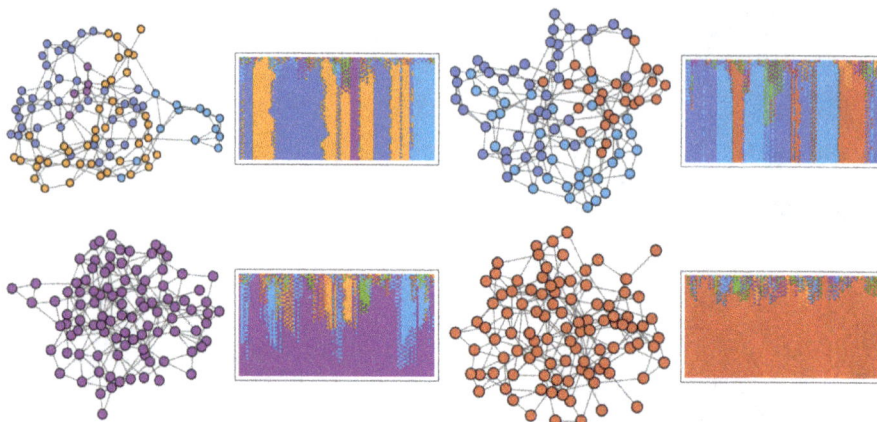

Exploration of Distributed Consensus on Graphs

MOHD HAMZA AHMAD

In this project, we studied distributed consensus in various distributed systems to gain a better understanding of its implications in real-world scenarios across different domains, such as social media, family and biological networks. We explored distributed consensus using different types of distributions, including Bernoulli, Watts–Strogatz and Barabási–Albert distributions. Additionally, we investigated the effects of parameters such as the number of colours and edge probability on reaching a consensus. Our approach began with the analysis of different distributions in preferential consensus scenarios. We explored the trends in reaching consensus by varying parameters. The project also delved into the practical interpretations that can be derived from the analysis and their importance in real-world scenarios. In future, I would like to further explore the Strogatz paradox, investigate the distributions using different iterative dynamics and test the interpretations on real-world data.

Scan or visit
wolfr.am/WSS2023-Ahmad

Hi Data: A Data Science Primer for High-School Students with Wolfram Language

HASAN KHAN

Data science is a fascinating field that utilizes scientific methods, processes, algorithms and systems to extract knowledge and insights from both structured and unstructured data. It's a discipline that involves a lot of creativity and critical thinking, and can be used to make data-driven decisions and predictions, making it an essential skill for the twenty-first century. This introductory course aims to familiarize high-school students with data science using Wolfram Language. Wolfram Language is a high-level programming language developed by Wolfram Research. Known for its sophisticated computational capabilities, it is widely used in scientific research, engineering, data science, machine learning and many other areas. In data science, Wolfram Language stands out because of its intuitive syntax, massive inbuilt computation and knowledge engine and extensive data visualization capabilities.

Scan or visit
wolfr.am/WSS2023-Khan

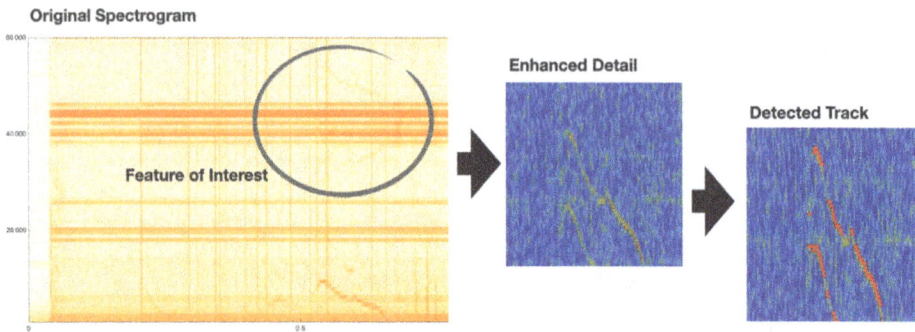

Original Spectrogram · Feature of Interest · Enhanced Detail · Detected Track

Frequency Track Detection in Polarimeter Data

JILL FOLEY

This project involved analysis of existing time series data. In some datasets, there exists a phenomenon of interest in which there is a component with a time-varying frequency. This signal is weak, with an amplitude near the level of background noise. The goal of the Wolfram Summer School project was twofold: first, to discern datasets in which the events of interest occur; and second, to determine the frequency track. A dataset previously identified to contain the phenomenon was studied in detail. First, a spectrogram was created that showed the feature, but it did not stand out. Optimization of the spectrogram was performed, resulting in images that clearly showed the phenomenon when present. This same approach was applied to additional datasets, and a classifier function was trained to distinguish the images displaying cases of the desired signal from those without. Additionally, a frequency track was identified by first detecting peaks in the spectrogram and then retaining only those that were found to have a suitable number of "neighbor" peaks, indicating a frequency track as opposed to random noise.

Scan or visit
wolfr.am/WSS2023-Foley

Exploration of Symmetric Nonlinear Wave Equations

KAMIL DUTKIEWICZ

This project investigates the solutions to the PDE
$\partial_t^2 u(t, x) = \partial_x^2 u(t, x) - a(u(t, x) + b)\,(u(t, x) - b)\,u(t, x).$ *Because of a nonstable equilibrium at $u = 0$, the system shows chaotic behavior. Interesting structures are observed in numerically computed $u(t, x)$ solutions, and their features are analyzed for different a and b parameters. A time-independent version of the previous equation is studied and solutions are used as initial conditions, resulting in nonstable equilibrium states.*

Scan or visit
wolfr.am/WSS2023-Dutkiewicz

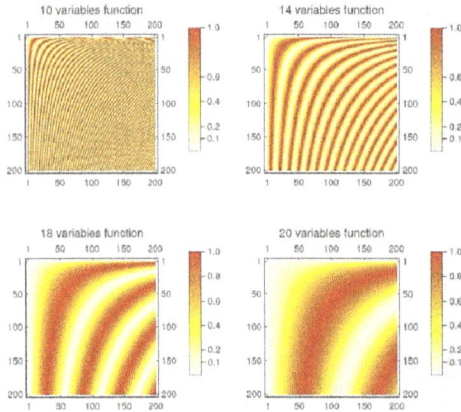

Quantum Gate Complexity of Boolean Functions

JAKUB GRABARCZYK

In the late 1990s, when famous quantum algorithms (especially Shor's) were published, the race to build the first quantum computer began. Occurring in that time were many technological difficulties—due to the lack of an obvious way to represent quantum states, the unstable nature of quantum states themselves and a complicated measurement process. One of these challenges is designing quantum circuits that perform functions fast and simply.

Scan or visit
wolfr.am/WSS2023-Grabarczyk

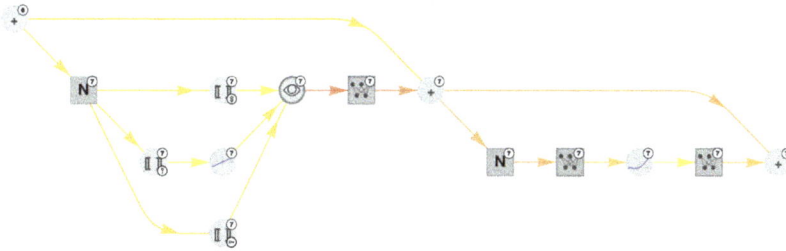

Unveiling the Enigma of GPT-2: Decoding Its Contextual Inference Mechanism

MEHRDAD MOHARRAMI

Goal: *Our study focuses on exploring GPT-2's capabilities in detecting and understanding contexts in text. We aim to determine whether context learning occurs gradually or if different components in GPT-2 are responsible for distinct context extraction. By delving into the inner workings of GPT-2, we aim to gain a better understanding of its ability to distinguish contexts.*

Result: *Through our probing technique, we have gained valuable insights into the context detection capabilities of GPT-2. Through numerical investigation, we have discovered that certain self-attention layers are likely responsible for specific types of context detection. Furthermore, we have observed that context detection occurs both gradually and abruptly within GPT-2. While the overall quality of the input signal tends to improve as it traverses the layers, there are notable instances where sudden jumps occur, with the pattern influenced by the specific context being analyzed.*

Future work: *To obtain a comprehensive understanding of GPT-2's inner workings, further exploration is needed. This includes studying each attention head independently, meticulously examining and fine-tuning the probe's design and conducting investigations on larger datasets to uncover the diverse roles of GPT-2's components. Additionally, the exploration of opposing sentences, differing by only a single word yet yielding distinct contexts, holds promise in unraveling the intricate circuits of GPT-2.*

Scan or visit
wolfr.am/WSS2023-Moharrami

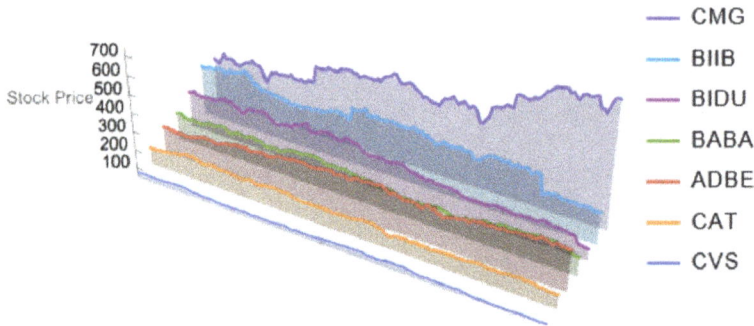

Forecasting Stock Prices Using LLMs with Financial News

VATSAL PIYUSH SHAH

Financial news articles play a crucial role in driving stock market trends, often triggering swift market reactions. By capitalizing on the speed of artificial intelligence in processing and reacting to these news articles, an advantage can be gained in navigating the stock market. This research investigates this premise, hypothesizing that the integration of news articles and stock price history can effectively predict market movements. Utilizing Wolfram Language and the large language models GPT-2 and BERT, we developed and trained a series of classifiers, leveraging the text from news headlines and historical stock prices as inputs. Our results indicate that the incorporation of financial news does indeed enhance the prediction accuracy, with the model that merges GPT-2 embeddings and stock price history delivering the best performance. Future work aims to fine-tune this approach by adjusting the historical data length and prediction horizon and incorporating bid-ask spreads, with the ultimate goal of creating a practical tool that can anticipate market trends swiftly and accurately based on the combined influence of financial news and stock price history.

Scan or visit
wolfr.am/WSS2023-Shah

Investigating the Claims of "Grover's Algorithm Offers No Quantum Advantage"

SURAJ VISHWANATH

The purpose of this project is to critically examine and evaluate the claims put forth in the paper titled "Grover's Algorithm Offers No Quantum Advantage." The paper challenges the widely accepted notion that Grover's algorithm, a quantum search algorithm, provides a significant speedup compared to classical algorithms. The project aims to investigate the validity of these claims and determine whether Grover's algorithm truly offers no quantum advantage.

Scan or visit
wolfr.am/WSS2023-Vishwanath

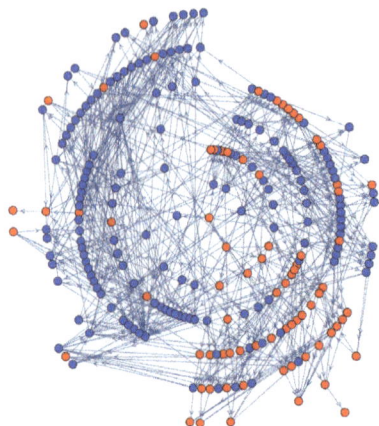

Hackenbush Games as Multicomputational Systems

ZSOMBOR ZOLTÁN MÉDER

Hackenbush is a class of two-player, zero-sum perfect information games. It is played on a graph whose edges are available to either or both players for removal. Players alternate in removing edges of the graph. After each removal, any edges unconnected to a distinguished vertex are removed from the graph. In the standard version of the game, the first player without any available moves loses. We explore Hackenbush games by constructing multiway graphs corresponding to all possible sequences of moves. Using this multiway graph, we determine the winning player at each position, and determine the probability for each player to win when they pick their moves randomly. Future work on Hackenbush can build on our framework to determine the (Conway-)value of Hackenbush graphs.

Scan or visit
wolfr.am/WSS2023-Meder

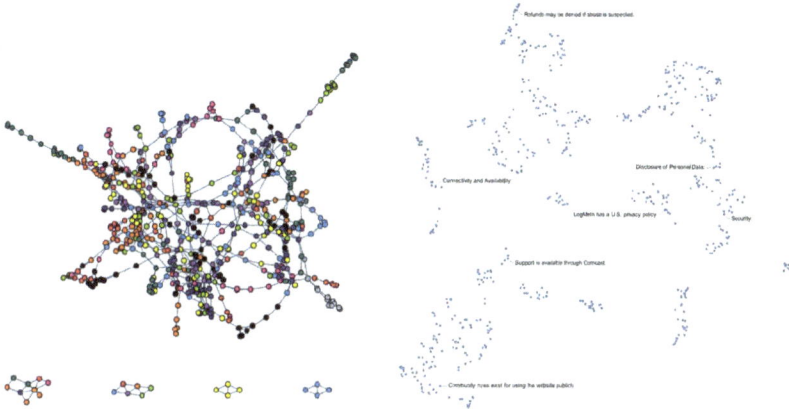

Unveiling the Veil: A Semantic Analysis and Visualization of Web Terms & Conditions

DONGMING JIN

This project seeks to shed light on the intricacies and obscurity of online terms and conditions (T&Cs) that users frequently agree to without comprehensive comprehension. We employed ChatGPT (LLM) and Wolfram Language to conduct a shallow semantic examination of 161 T&C documents from 70 different companies and established a web service for on-demand T&C analysis. We discovered that while the LLM's capabilities are hampered by token limits and the quality of prompts, it holds tremendous promise for tackling intricate text-processing tasks that demand extensive domain expertise. We plan to further harness this capability, fostering a community-driven movement reminiscent of the open-source revolution.

Scan or visit
wolfr.am/WSS2023-Jin

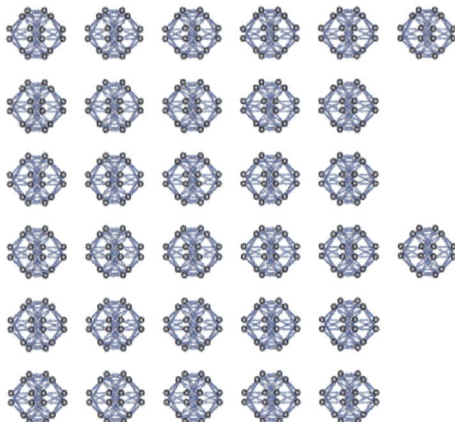

Multiway Systems for Exactly Solved Lattice Models

AISEL SHIRALIEVA

In this project, we build a multiway graph for the two-dimensional Ising model. We compute the energies for each configuration. We try different rules to generate configurations of the system, which creates new multiway graphs. We look at loops in a multiway graph; given an evolution function, some states can only reach some of the other states, and in general there would be multiple connected components in the graph, loops, stable states, etc. This project can be extended to the Potts model.

Scan or visit
wolfr.am/WSS2023-Shiralieva

Exploring Minimized Reversible Circuits for *n*-Bit Functions

BENJAMIN PETER

Reversibility is an important trait for logic systems and an essential aspect of quantum computation. The conversion of traditional Boolean gates such as AND and XOR to the universal Toffoli gate allows for reversibility in classical n-bit–to–n-bit systems. Using logic pattern recognition and Mathematica's powerful Boolean functionality, gate counts for reversible circuits can be determined from a given input function.

Scan or visit
wolfr.am/WSS2023-Peter

General Framework for Agent-Based Modeling

MORITZ KALHÖFER-KÖCHLING

Ranging from predator–prey systems to trade markets, automobile traffic, thermodynamics, fluid dynamics and so on, agent-based models (ABM) encompass a wide variety of applications. Seemingly trivial update rules (~equations of motion) yield a plethora of unexpected complexity, such as swarm synchronization, phase transitions or the spread of diseases.

Thus it comes as no surprise that many programming languages tailored to the execution of ABMs have been developed since their inception in the late 1940s. While plenty of different adaptions of ABMs for Mathematica have been explored, a concise and intuitive implementation has yet to be shaped. The aim of this work is to grant a simple agent simulation function encompassing both predefined models and customizability, hosting many of the aforementioned applications.

In the following, the two implemented models (predator–prey and Conway's Game of Life) are laid out in short, followed by two explanations on how to create different models building on the existing paradigm.

Scan or visit
wolfr.am/WSS2023-Kalhofer-Kochling

Exploring Facade Patterns in Urban-Built Environments: A Step toward Circular Architecture

BING YANG

This project explores patterns in the facades of buildings in urban environments using image-clustering techniques, with the potential for further exploration into component segmentation. The aim is to identify common facade types and investigate their distribution across different neighborhoods and time periods. The insights gained from this project will contribute to the field of circular architecture by providing a better understanding of the existing built environment. By identifying common facade types and their distribution, we can start to view the built environment as a resource bank, where building elements such as windows, doors, panels, etc. can be reused or recycled, thus promoting a more sustainable and circular approach to urban planning and architectural design.

Scan or visit
wolfr.am/WSS2023-Yang

Viral Dynamics Modeling for Hepatitis B Virus (HBV) and Hepatitis Delta Virus (HDV)

DANNY BARASH

Hepatitis B virus (HBV) and hepatitis delta virus (HDV) coinfection is considered the most severe form of chronic viral hepatitis. Drugs against HDV are under development, which serves as motivation to model the interplay between HBV and HDV under anti-HDV treatment. We first examine a simple differential equation model for studying the coinfection at the chronic stage, solving the model equations and performing the fitting. We offer an interactive user interface for experimenting with the model parameters. We then examine the HBV acute infection stage and model the dynamics of the cell-to-cell spread by cellular automata. We find out that a simple cellular automaton rule is able to capture a biphasic behavior in the HBV cell-to-cell spread of the infection. As future work, cellular automata can be extended to model more features in the dynamics of the HBV acute infection and study its multi-phases that have been observed.

Scan or visit
wolfr.am/WSS2023-Barash

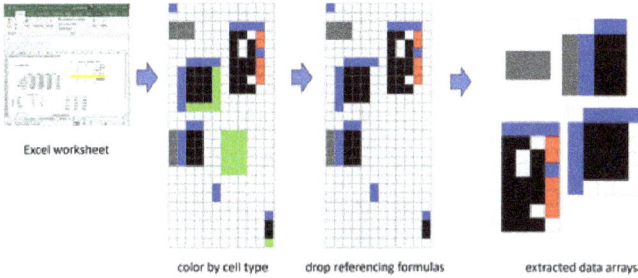

Excel worksheet

color by cell type   drop referencing formulas   extracted data arrays

Extracting Rectangular Data Arrays from Microsoft Excel Files

DAVID DEBROTA

Exploration of data commonly involves statistics and visualization tools that operate best on rectangular arrays of cells containing numbers, strings and/or dates/times, with minimal missingness. In this project, "intelligent," easy-to-use tool functions will be developed to operate on Microsoft Excel files, isolate and extract rectangular arrays of data, capture accompanying metadata, and help the user understand any missingness that is present.

Scan or visit
wolfr.am/WSS2023-DeBrota

**WOLFRAM
+ AI for Education**

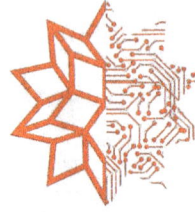

Fusing Prompt Engineering and Wolfram Coding for Enhanced Tutoring

GHASSANE ANIBA

This project aims to advance interactive tutoring by leveraging the capabilities of large language models (LLMs), thereby enhancing the learning experience for students. The proposed research integrates prompt engineering techniques, utilizing LLMs like GPT-3.5 and GPT-4, with the versatility of Wolfram code programming to create comprehensive subject-sheets on any requested topic. Despite the progress made with LLMs such as GPT-4, challenges remain in generating complex codes, including interactive animations, as well as in ensuring the inclusion of accurate and up-to-date scientific references. To address these issues, this project developed a system that generates a "subject-sheet" comprising an abstract, keywords, an introduction, interactive animations and current scientific journal and book references relevant to each user-requested subject. To achieve this, we employed LLMs to generate textual content and the Wolfram Demonstration Project webpage from the Wolfram website for the creation of animations. The most relevant Demonstrations were selected using LLMs. References were obtained through a service connection between Wolfram and the Crossref website. All collected data was curated and consolidated into a single generated notebook based on a predefined template.

Scan or visit
wolfr.am/WSS2023-Aniba

Exploring Spin Dynamics: An Ising Model Study

HENGYUAN XU

This project revolves around simulating and exploring the Ising model, with a focus on studying its behavior and properties. We utilize the Metropolis–Hastings algorithm to obtain the stationary distribution by evaluating the Hamiltonian. By conducting Markov chain Monte Carlo (MCMC) simulations, we calculate the heat capacity, magnetization and susceptibility of the model to analyze the phase transition phenomena within the 2D Ising model. Additionally, we investigate the impact of various parameters like temperature and interaction coefficients on the patterns observed in a two-dimensional setting. Moving forward, our plans involve applying the Ising model to a fully connected graph, incorporating mean-field theory and using this framework to simulate complex systems such as spin-glass and melting crystals.

Scan or visit
wolfr.am/WSS2023-Xu

Coffee Roasting Curve Analysis

IVAN EDGARD PRADO LÓPEZ

The coffee roasting curve is a fundamental component in the coffee production process. The detailed analysis of this curve provides valuable information about the quality and roasting profile of the resulting coffee. However, this analysis can be complex and requires a deep understanding of the data collected during roasting, leading to a prolonged duration for the analysis of multiple curves.

The combination of Wolfram Language and GPT models offers a powerful tool for analyzing the coffee roasting curve. By utilizing these technologies, roasters can gain a deeper understanding of the roasting processes in less time, identify improvement opportunities and make informed decisions to produce a well-developed roasting profile, resulting in high-quality coffee with exceptional flavor.

Scan or visit
wolfr.am/WSS2023-Lopez

Empirical Metamathematics: Extending the Lean-to-Mathematica Bridge

JACK HESELTINE

The goal of this project is to do some metamathematics. The first step will be to work on a Wolfram Language importer for the Lean theorem prover, a currently popular system that blends interactive and automated theorem proving, two slightly different approaches to finding proofs. Most importantly, Lean draws on the Lean mathlib collection, made up of thousands of theorems and definitions.

Quick and automated proof checking can help anyone who might like to outsource their proving! Mathematica actually has proving capability and the ProofObject data structure for equational logic; Lean goes a little further by also allowing quantified terms.

An institute at my university (Research Institute for Symbolic Computation at Johannes Kepler University, Linz, Austria) is deleveoping a theorem prover that plugs into Mathematica. Bringing in big community math theorem packages could be a good way to advance this project. But what is in such a package, and how can I explore it? This brings me to metamathematics and exploring Lean and mathlib using these approaches.

Scan or visit
wolfr.am/WSS2023-Heseltine

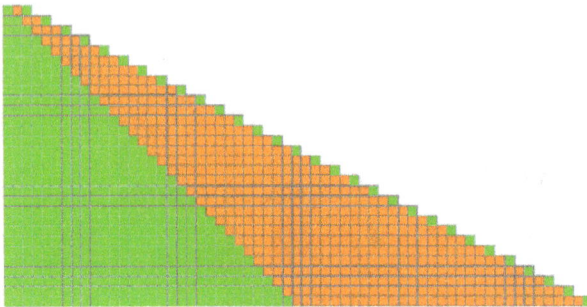

Exploring Neighbor-Dependent Substitution Systems Using Cluster-Size Entropy and MSD

JIM TORTELLA

There exists a notable interdisciplinary interest in deciphering and characterizing the dynamics and emergent phenomena found within biological media. In this context, lattice models are often utilized to enrich our understanding of these systems [1]. Bearing in mind that these media are generally modeled out of equilibrium and that various diffusive behaviors have been identified [2], this study delves into the exploration of the peculiarities presented by substitution models and considers how it might be adapted to tackle similar problems.

In the present work, we use Wolfram Language, cluster-size entropy and mean square displacement to shed light on certain aspects of the dynamics of substitution models. We find that the inherent variability in system size, a fundamental characteristic of substitution systems, impacts our entropy measurements. Even so, cluster-size entropy serves as a valuable guide, providing clues about the emerging patterns within the system. Additionally, we find that the system size expands following a pattern of super diffusion.

Although we believe these characteristics may hold relevance for future applications, there remains much to uncover regarding how one might model and understand a biological system using substitution systems.

Scan or visit
wolfr.am/WSS2023-Tortella

```
‹|Game 1 → ‹|Alice Play → ROCK, Bob Play → ROCK, Winner → Draw|›,
 Game 2 → ‹|Alice Play → PAPER, Bob Play → ROCK, Winner → Alice|›,
 Game 3 → ‹|Alice Play → SCISSORS, Bob Play → PAPER, Winner → Alice|›,
 Game 4 → ‹|Alice Play → ROCK, Bob Play → ROCK, Winner → Draw|›,
 Game 5 → ‹|Alice Play → PAPER, Bob Play → ROCK, Winner → Alice|›,
 Game 6 → ‹|Alice Play → SCISSORS, Bob Play → ROCK, Winner → Bob|› |›
```

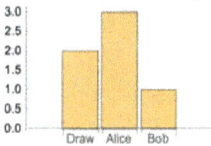

GPT + Wolfram Language = End of Days

JONATHAN MILLER

Can an LLM predict what I, or another LLM, will do next? We study the simplest zero-sum game, "matching pennies." Agents Alice and Bob simultaneously select a 1 or a 0. If they both select the same value, then Alice wins; otherwise, Bob wins. Alice is an LLM (GPT-3.5 or GPT-4, say), and Bob is a human or another LLM. The agent who, based on past games, can most reliably predict what the other agent will propose in the future, will win the most games. Humanity doesn't get to choose the game, which is arguably the real-world game of life.

We create and execute examples of a human playing the game with GPT-3.5 and GPT-4, and then some examples of GPTs playing with each other.

Scan or visit
wolfr.am/WSS2023-Miller

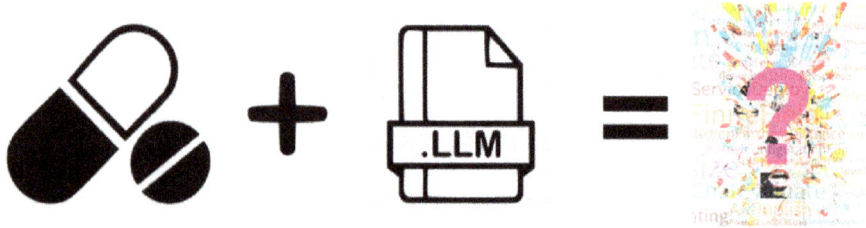

What Happens if You "Drug" LLMs?

MICHAEL GREY

In a project defined by Stephen Wolfram, we investigated the simulation of "drugging" effects on large language models (LLMs). The initial idea, inspired by neuroscience, was to adjust LLM connectivity to mimic the effects of LSD on the brain. However, due to time constraints and the complexity of accurately simulating psychedelic effects, we altered our approach. Following Theodore Gray's suggestion, we started by manipulating the LLM's temperature to induce a "hallucination effect" on specific text strings, aiming to abstract language. We developed a function for ChatGPT to interpret this altered text, testing its ability to detect its own temperature changes. This set the stage for the concept of "drugging" language models. We also created embeddings for generated sequences to visualize meaning space to predict temperature and developed a psycholinguistic function to check text coherence. Our findings suggest that LLMs' coherence is greater when we have a meaningful psycholinguistic scale. Next steps include refining the coherence function, submitting it to the Wolfram Function Repository and further exploring xenolinguistic models to develop deeper and more meaningful psycholinguistic functions.

Scan or visit
wolfr.am/WSS2023-Grey

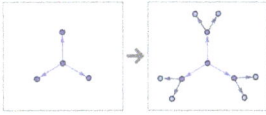

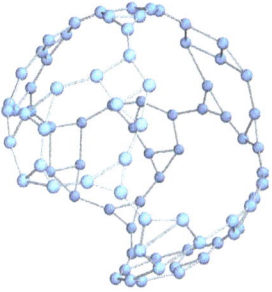

Studying the Behavior of Simple Network Rewriting Systems

MINAYA ALLAHVERDIYEVA

The aim of this project is to explore the trivalent networks generated by random replacements using rewriting rules from the Wolfram model. The goal is to connect this to φ^3 scalar field theory. In the limit of large n, the SU(n) internal symmetry is dominated by Feynman diagrams that exhibit the structure of planar trivalent networks. Hence, the enumeration of trivalent planar graphs play a crucial role in understanding Feynman diagrams for such theory. By applying simple rules iteratively, we can generate complex network structures and examine their emergent behavior in terms of network connectivity and average statistical properties. This research contributes to a deeper understanding of the φ^3 theory and provides insights into the dynamics of these computational systems.

Scan or visit
wolfr.am/WSS2023-Allahverdiyeva

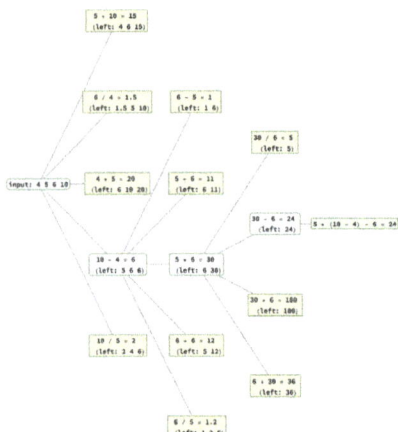

Tree of Thoughts: Exploratory Problem Solving with LLMs

ROBERT GIOMETTI

Language models have proven effective problem solvers across a wide range of problems in many domains, but are limited by a token-level decision-making process during inference. Because of this, initial decisions play a pivotal role and early mistakes can prevent success in later decision making. Tree-of-Thought (ToT) is a decision-making framework introduced in that generalizes Chain-of-Thought prompting, enabling language models to explore branching possibilities with coherent units of text (thoughts) as an intermediate step. ToT allows LMs to perform deliberate decision making by considering multiple different reasoning paths and self-evaluating choices to decide the next course of action, as well as looking ahead or backtracking when necessary to make global choices. In this project, we implement the ToT framework in Wolfram Language for the first time, reproduce a solution for the game of 24 and begin to explore the use of this framework for more general problem solving in the context of coding challenges.

Scan or visit
wolfr.am/WSS2023-Giometti

Exploring Generalized Collatz Functions

RUSSELL MARTINEZ

The Collatz conjecture, or 3x + 1 problem, is an unsolved mathematics problem infamous for being easy to explain but remarkably difficult to solve. The Collatz function relies on choosing a positive integer and continuously repeating the following operations: if the number is even, divide it by two; if the number is odd, multiply it by three and add 1. The conjecture states that every positive integer will always reach one. This project uses Wolfram Language to explore established and potentially new methods of visualizing generalized Collatz functions given by a defined U-function. The numerous phenomena observed include structured distributions of prime number counts in iteration graphs, patterns in digit plots of different bases and manipulated trajectories using truncated p-adic integers. The observations in the project will hopefully allow others to launch further investigations. This project attempts to visualize the implications of generalized Collatz functions in number theory and geometry from an experimental perspective. An introduction to some representations of generalized Collatz functions precedes the explorations relevant to the aforementioned fields. The treatments described in this project can apply to other generalized Collatz functions. More rigorous analysis is required to appreciate the observations made.

Scan or visit
wolfr.am/WSS2023-Martinez

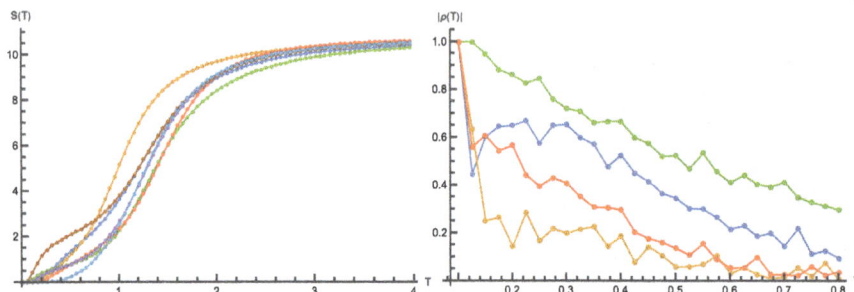

Study of the Possibility of Phase Transitions in LLMs

SEBASTIÁN BAHAMONDES

In this project, we quantitatively analyzed the behaviour of the output of the GPT-2 LLM and its dependence on the temperature parameter that is used in the final sofmax layer of probabilities. It has been noticed that as temperature decreases toward absolute zero, the text output of neural networks, such as GPT-3.5 and GPT-4, tends to become completely deterministic and uncreative, while as temperature increases arbitrarily, the sentences that are output by the LLM become chaotic and nonsensical. Therefore, the question of the existence of a critical temperature at which LLMs exhibit an abrupt transition between these two types of behaviour arises, as well as the question of this potential critical temperature being the hallmark of a phase transition–like process in LLMs. The hypothesis of this project is that there is a critical temperature at which some set of quantitative parameters associated to LLMs changes abruptly, or shows some noticeable qualitative change. The main goal of this project is therefore to identify an appropriate quantitative parameter associatede to the output of an LLM and see how it behaves at increasingly low temperatures.

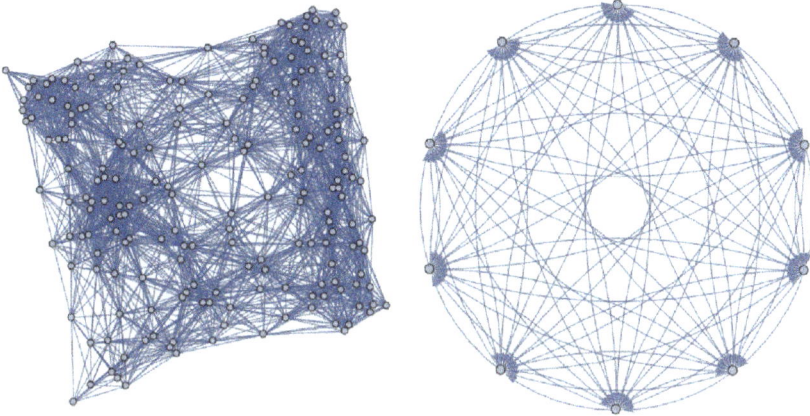

Exploring the AdS/CFT Correspondence in Discrete Spacetimes via Tensor Networks

SRISHTI NAUTIYAL

The AdS/CFT correspondence posits an equivalence between a gravitational theory (d + 2-dimensional anti-de Sitter spaces) and a lower-dimensional non-gravitational theory (d + 1-dimensional conformal field theory). This duality is solely dependent on the AdS boundary conditions. This project seeks to model the AdS/CFT duality within the Wolfram framework by exploring the effect of the AdS boundary conditions on bulk properties such as dimensionality and curvature. We use causal graphs obtained by sprinkling to represent our bulk space. These graphs have coordinates associated with each vertex; these coordinates are utilized to demarcate subgraphs forming the "bulk" and "boundary" of this space. We define a tensor network on the boundary to calculate the entanglement entropy and then explore connections to the bulk calculations like the Hausdorff dimensionality and Ricci scalar curvature.

Scan or visit
wolfr.am/WSS2023-Nautiyal

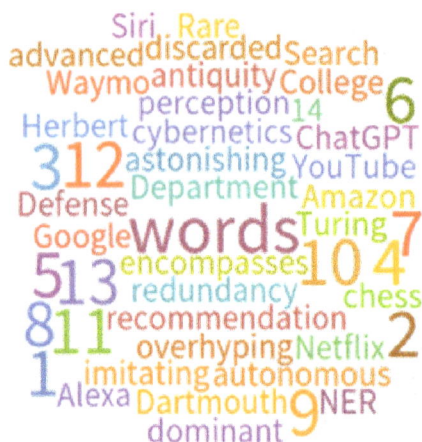

Analysing Rare and NER Words in Wikipedia

TAWSIF AHMED

In this experiment, we're going to analyse rare and named entity recognition (NER) words in Wikipedia articles using large language models through LLM functions in Wolfram Language.

Scan or visit
wolfr.am/WSS2023-Ahmed

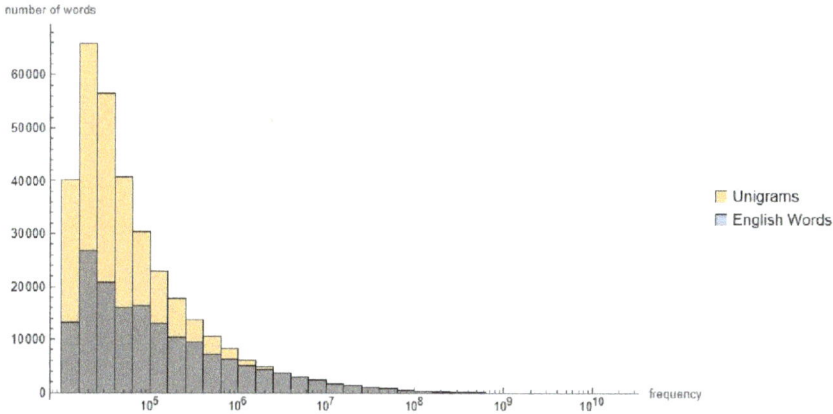

Investigating the Extent of the English Language

THIAGO HAN

This project aims to dive into the challenge of counting the total number of words in the English language, which is a complex task given the ambiguities of the definition of a word. It attempts to get an estimate by employing the LLM features of Wolfram Language to filter out real words, reach a final count and visualize words' extent based on their frequency of use.

Scan or visit
wolfr.am/WSS2023-Han

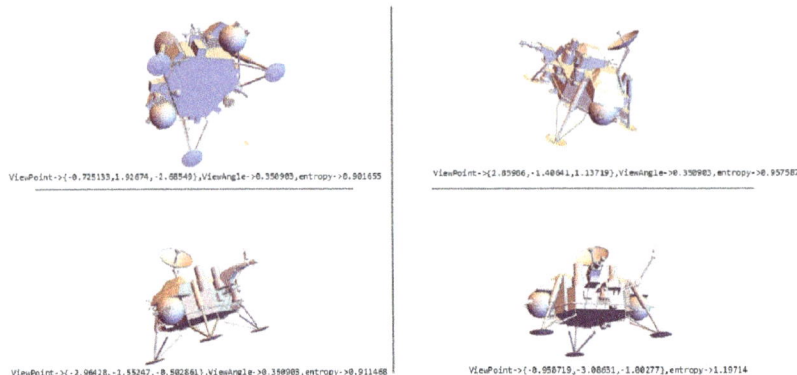

ViewPoint->{-0.725133,1.92674,-2.68549},ViewAngle->0.350903,entropy->0.901655

ViewPoint->{2.83986,-1.40641,1.13719},ViewAngle->0.350903,entropy->0.957587

ViewPoint->{-2.96428,-1.55247,-0.502861},ViewAngle->0.350903,entropy->0.911468

ViewPoint->{-0.958719,-3.08631,-1.80277},entropy->1.19714

Quantifying an Optimal Image to Maximize Information for 3D Reconstruction

TIYASA SARKAR

Reconstructing 3D data from a single viewpoint in a fundamental human vision functionality is one of the most challenging tasks that most computer vision algorithms struggle with. In this project, we try to quantify and evaluate information content through entropy measure such that we obtain an optimal image, which could provide the maximum information for 3D reconstruction if we are constrained to make a reconstruction given one image. We try to explore if the entropy-based calculations could be a good metric to reduce the number of images for image reconstruction. We calculated joint entropy between the difference of grayscale pixel intensities in horizontal and vertical directions of images of a 3D object at different viewing positions and angles, and choose the image with maximal entropy.

Scan or visit
wolfr.am/WSS2023-Sarkar

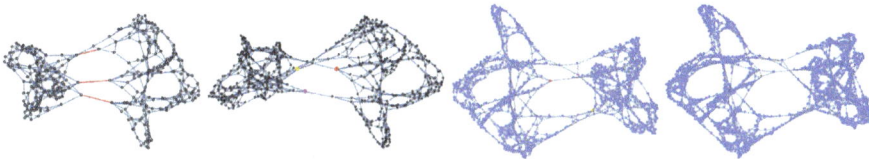

Domain Walls in Graphs

TOMASZ MAZUR

Domain walls are soliton topological defects believed to have occurred in the early universe during phase transitions. These objects separate patches of space with different vacuum expectation values resulting from the spontaneously broken symmetry of an underlying potential. Domain walls contribute significantly to the energy density of the universe and must be considered when constructing cosmological models involving spontaneous symmetry busting (SSB). In this project, our aim is to explore the possibility of similar behavior in graphs. We will implement methods for merging different graphs with wall-like structures and attempt to create an algorithm capable of finding them. Additionally, we will investigate the behavior of introduced domain walls while evolving the merged graphs.

Scan or visit
wolfr.am/WSS2023-Mazur

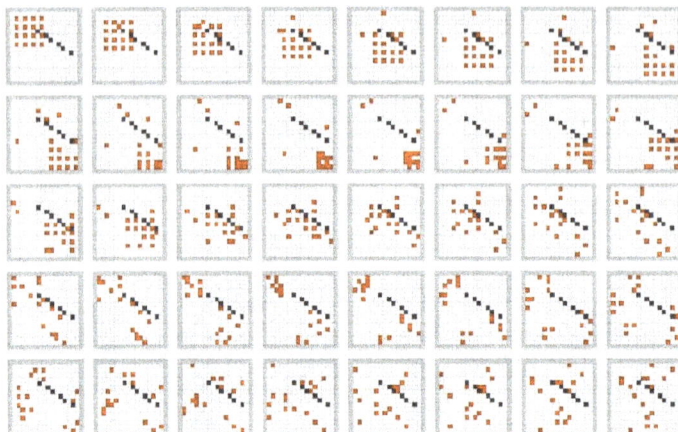

Study of Statistical Evolution of a Gaseous System as a Consequence of Iterative Simple Rules

UTKARSH PATEL

This project aims to analyze a simple 2D gas-like system in a container and look for the emergent properties of random behaviour in its constituent particles given specific initial conditions and cellular automaton evolution rules. These conditions can range from the introduction of a single obstruction cell or a group of obstruction cells placed at specific coordinate locations inside the container; these cells interfere with the gas constituents to change the initial number density, position or orientation of the gaseous constituents. This leaves a scope to evaluate and study a lot of inter-relationships between various measurable properties of the evolved system. The measurement of randomness and the system's ergodicity as a function of the perturbation strength and relative positioning, time/ensemble averages and autocorrelations of quantities like particle density distributions, and cyclicity within the system as a function of time are a few of those properties.

Scan or visit
wolfr.am/WSS2023-Patel

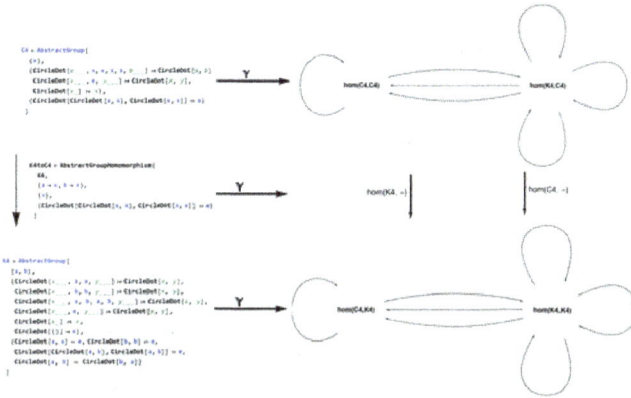

Toward a Homoiconic Foundation in Wolfram Language

ADAM MILLAR

This project explores preliminary work in creating a homoiconic foundation of mathematics in Wolfram Language. A homoiconic foundation is one where the connection between the syntax and semantics of a formal theory is established from the beginning. The basis for this work is inspired by the general methodology of mathematical logic, where systems of derivation rules (proof theory) and systems of models (model theory) are developed independently and then connected via an interpretation. Constructing this interpretation is not always a straightforward process. A homoiconic foundation would put the effort of constructing the interpretation into the building blocks of the theory. This project presents a syntactic formal theory for two small groups in terms of Wolfram Language rewrite rules and data structures. It then expresses the categorical structure of the theory in terms of the categorical tools from Wolfram Language. The resulting functor is a component of a larger interpretation. There is a great deal of possible future work to perform. The functions and techniques used in this project are exploratory and preliminary, and refinement is called for in many places. Beyond that, the methodology of the project needs to be generalized to other formal theories so that it can legitimately serve as a foundation.

Scan or visit
wolfr.am/WSS2023-Millar

Investigating LLM-Agent Interactions

ADRIAN MLADENIĆ GROBELNIK

Our project hypothesizes that complex emergent behaviors can arise from multi-agent simulations involving large language models (LLMs), potentially replicating intricate societal structures. This was put to the test through three progressively more complex simulations, where we evaluated the agents' understanding, task execution and capacity for strategic interactions such as deception. Our observations suggest that while some LLMs, like GPT-3.5, may encounter difficulties with certain tasks compared to GPT-4, they effectively contribute to iteratively updating complex simulation environments with pertinent data. These findings highlight the challenges in discerning the extent of agent understanding and their capability limits. The next phase of our research will intensify the sophistication of the agent architecture, enhance the complexity of the simulations and conduct larger-scale experiments with more iterations to further delve into the dynamics of LLM-agent societies.

Scan or visit
wolfr.am/WSS2023-Grobelnik

Cycles in $x \to (10x + 1) \bmod i \{i, 1, 10\}$ Rule

Graph	Number of Cycles	Index
	4	{1, 2, 5, 10}
	2	{3, 6}
	1	{4}
	1	{7}
	1	{8}
	1	{9}

Explorations on the Ruliad via Modular Arithmetic

JUAN ARTURO SILVA-ORDAZ

We explore graph properties of the ruliad by changing parameters of modular arithmetic functions. We explored simple rules applying the modular operation on linear, quadratic and cubic functions. We found isomorphic graphs and cycles of many sizes for every function tried. A general framework to study the isomorphisms and cycle size needs to be built to further study and properly classify the structure of these rules.

Scan or visit
wolfr.am/WSS2023-Silva-Ordaz

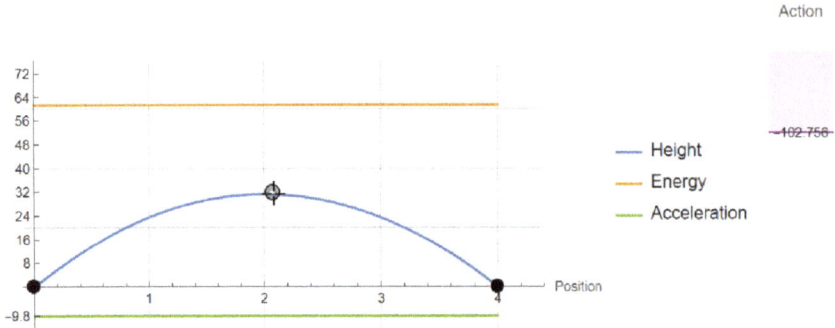

Interactive Demonstration of Lagrangian Mechanics

ISAÍAS MARTÍN MUÑOZ CUBERO

The main idea of this project is to approach Lagrangian mechanics in a way people with basic knowledge can understand it and visualize it better using Wolfram Language. The goal is to make an interactive game in which one can discover the minimum of the action, which is the integral of the Lagrangian, simply by changing parameters in trial solutions to the problem.

Scan or visit
wolfr.am/WSS2023-Cubero

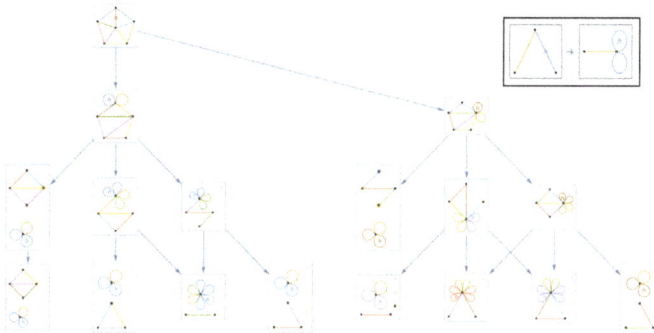

The Ruliology of Network Mobile Automata

JOEL CHOI

Network mobile automata (NMA) are systems in which clusters of nodes in a hypergraph are transformed based on rules that not only change the shape of the cluster but also assign a tag to one particular edge and move it around as the system evolves. The labeled hyperedge is analogous to a mobile head, like in a Turing machine, and by testing various initial conditions and rules characterized by hypergraph rule signatures, we strive to understand the evolution behavior of the NMA. This project revealed early signs of complex behavior, and by optimizing the process for checking hypergraph isomorphisms, we may be able to find even more unpredictable evolution. Also, there is more investigation to be done to characterize planarity-preserving rules, which preserve dangling edges, and the types of evolutions they would lead to.

Scan or visit
wolfr.am/WSS2023-Choi

Exploration and Characterization of Mobile Automata

LAP SUM CHAN

In this project, we attempt to characterize mobile automata in the form of network graphs. By applying dimension-reduction techniques, we transform these high-dimensional network graphs into a simplified 2D space. Our findings reveal distinct clusters of rules, each exhibiting unique behaviors, thus opening avenues for further exploration and classification within the fascinating domain of mobile automata.

Scan or visit
wolfr.am/WSS2023-Chan

— Abs value of **Single** Angle Pair Bell
— Quantum Mechanics: Cos[2k]

Single-Photon, Single-Angle Pair-Bell Experiment

MARK MERNER

An initial Single Photon, single-photon, single-angle pair-based Bell-type experiment was conducted and completed as a proof of concept, sanity test and infrastructure preparation for final publication. The possibility of such an experiment emerged after generating six Bell aspect–type diagrams along with six absolute-value diagrams on a single page. At first, there was some doubt that it would be possible without a 'real' lab with an expensive spontaneous parametric down-conversion (SPDC)–entangled correlated source and automated time-stamped single-photon counters. However, after some preliminary test results, the experiment looked possible and was executed in a home optics setup with a simple power meter and an ordinary helium–neon laser. The intent now is to drive up sigma and turn this into a published experiment.

In parallel to this experiment, and motivated by two recent Bell-type entanglement of entanglement experiments which 'proved' that quantum mechanics requires complex numbers, preparation of the mathematical infrastructure for **single photon–based** *experiments of this type were made. As a follow-on to the completion of this experiment, the hope is to be able to do these single photon–type entanglement of entanglement experiments and show the value of the Mathematica QuantumTo⁚. MultiwaySystem-generated multiway diagrams in exploring the subtler aspects of quantum mechanics. For this part of the project, dozens of related multiway diagrams were generated using 10 common quantum operators entangled with single-pair Bell measurements. (Here, again, the value is to show that the big picture with a wide overview generates new possibilities.)*

Scan or visit
wolfr.am/WSS2023-Merner

RemoteSensing Paclet: GIBS Satellite Imagery and AρρEEARS Geographic Data Products

This project aims to develop robust tools for the integration of GIBS satellite imagery and AρρEEARS geographic data products into Wolfram Language. The focus of this project is to build efficient and flexible methods to import, visualize and interpret this geographic data using Wolfram Language as well as examples and documentation detailing the methods involved. We are delighted to announce the release of the RemoteSensing paclet to the Wolfram Language Paclet Repository as the culmination of these three weeks of work at the Wolfram Summer School. We will continue to work on this repository after the Summer School, and look forward to your feedback!

Introduction to the RemoteSensing Paclet

The RemoteSensing paclet contains various functions that allow you to make requests to the GIBS and AρρEEARS APIs for geographic products from various US federal agencies and work with the returned data in Wolfram Language. The RemoteSensing GIBS functions can be used with GeoImage and GeoGraphics through the GeoServer option, while the AρρEEARS functions can be used to request scientific-grade geographic data product samples.

We could go on at length explaining why this is useful, but why bother when we can simply mesmerize you with nice pictures? In the following sections, we will see some simple examples of what the RemoteSensing paclet can do. We hope our examples will inspire you to try the paclet for yourself, and perhaps even work with it in your own projects!

Installing and Loading the RemoteSensing Paclet

Before you get started with the RemoteSensing paclet, you'll need to install and import the paclet in your notebook:

In[∘]:= **PacletInstall["PhileasDazeleyGaist/RemoteSensing", UpdatePacletSites → True];**

In[∘]:= **PacletObject["PhileasDazeleyGaist/RemoteSensing"]**

Out[∘]= PacletObject[⬚ 🧊 Name: PhileasDazeleyGaist /RemoteSensing Version: **1.0.3**]

In[∘]:= **Needs["PhileasDazeleyGaist`RemoteSensing`"]**

Perfect—now for the good stuff!

GIBS Examples

A Few Words about GIBS

The NASA GIBS (Global Imagery Browse Services) API is a web service provided by NASA used to access and retrieve satellite imagery and related data. The API provides an interface to access a vast collection of global Earth science data through Web Map Tile Service (WMTS). Conveniently, it does not require authentication to access.

GIBS provides raster tiles, which can be useful for visualization and data exploration, but should be avoided for serious research purposes, as it does not provide access to the raw layer data. Its main advantages are that it is fast, pretty robust and very flexible.

GIBSData: Layer Metadata from GIBS

Use the GIBSData function to get information about the available GIBS geographic products.

Show available GIBS layers:

In[∘]:= **Take[GIBSData[], 5] // Column**

Out[∘]= Agricultural_Lands_Croplands_2000
Agricultural_Lands_Pastures_2000
AIRS_L2_Carbon_Monoxide_500hPa_Volume_Mixing_Ratio_Day
AIRS_L2_Carbon_Monoxide_500hPa_Volume_Mixing_Ratio_Night
AIRS_L2_Cloud_Top_Height_Day

As you can see, there are many layers to choose from:

In[ ]:= **GIBSData[] // Length**

Out[ ]= 1010

Get information about a layer:

In[ ]:= **GIBSData["Agricultural_Lands_Croplands_2000"] // Dataset**
(*GIBSData["Agricultural_Lands_Croplands_2000",All]//Dataset*)
(*GIBSData["Agricultural_Lands_Croplands_2000","Dataset"]*)

Out[ ]=

Title	Croplands (Global Agricultural Lands, 2000)
TileMatrixSet	GoogleMapsCompatible_Level7
Template	TemplateObject[Parameters: "TileCol", "TileMatrix", "TileMatrixSet", "TileRow"]
DateRange	—
DateDefault	—

Get the value of a specific property of a layer:

In[ ]:= **(*Possible values are:**
{"Title","TileMatrixSet","Template","DateRange","DateDefault"}*)
GIBSData["Agricultural_Lands_Croplands_2000", "Title"]

Out[ ]= Croplands (Global Agricultural Lands, 2000)

GIBSGeoServer: Requesting Map Tiles from GIBS

Use the GIBSGeoServer function to make a specified GIBS layer available to GeoGraphics and GeoImage.

Plot a **GeoGraphics** object, with a specified GIBS layer:

In[ ]:= **GeoGraphics[**
GeoServer → GIBSGeoServer["AIRS_L3_Surface_Air_Temperature_Monthly_Day"]]

Out[ ]=

GIBSGeoServer supports functions like GeoProjection:

In[ ]:=   **GeoGraphics[GeoServer → GIBSGeoServer["MERRA2_Surface_Albedo_Monthly"],**
     GeoProjection → #] &/@
   {"EqualEarth", "Orthographic", "PeirceQuincuncial", "ConicEquidistant"}

Out[ ]=

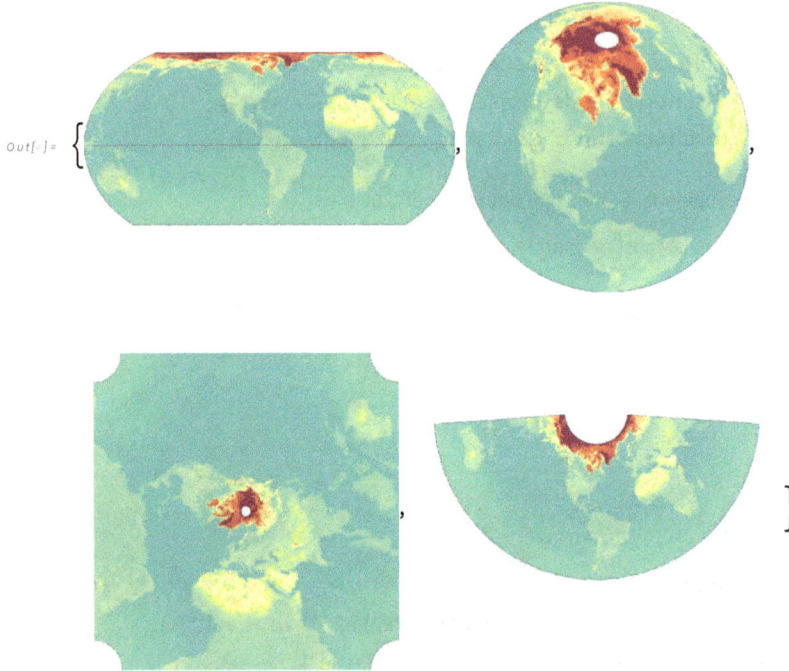

Combine two GIBS layers using **GeoGraphics**:

In[ ]:=   **GeoGraphics[{**
     GeoStyling["StreetMap", GeoServer →
       GIBSGeoServer["GPW_Population_Density_2015"], EdgeForm[Black]],
     Polygon[Entity["Country", "World"]]},
   GeoServer →
     GIBSGeoServer["AIRS_L3_Surface_Air_Temperature_Monthly_Day"], ⋯ ⊕ **]**

Out[ ]=

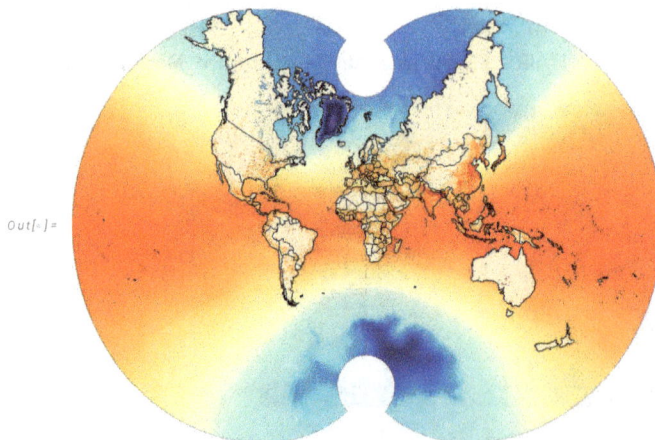

Use GIBSGeoServer in GeoImage:

```
In[ ]:= GeoImage["World",
         GeoServer → GIBSGeoServer["VIIRS_CrIS_NOAA20_BT_Band33_Fusion_Night"],
         GeoZoomLevel → 2, ImageSize → Large]
```

Out[ ]=

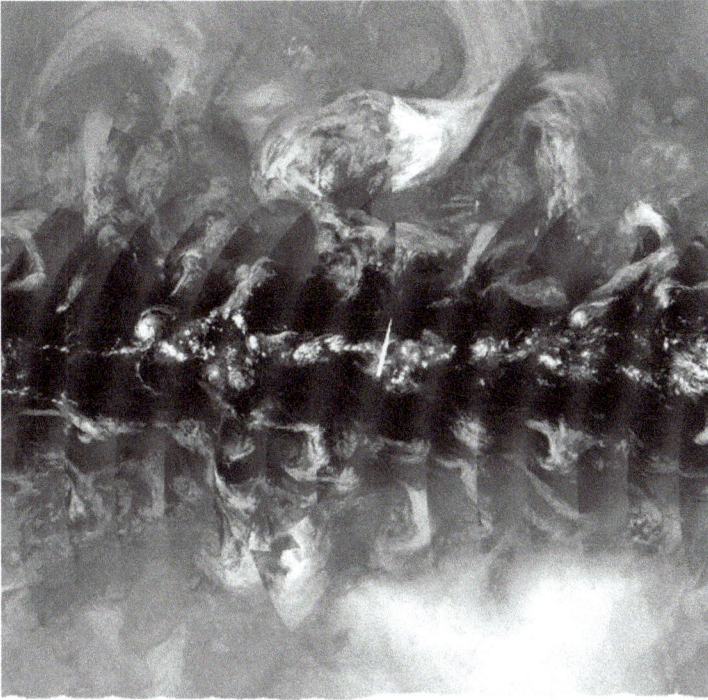

AρρEEARS Examples

A Few Words about AρρEEARS

AρρEEARS is a geographic data access service provided by NASA.

The AρρEEARS website describes it like so: "The Application for Extracting and Exploring Analysis Ready Samples (AρρEEARS) offers a simple and efficient way to access and transform geospatial data from a variety of federal data archives. AρρEEARS enables users to subset geospatial datasets using spatial, temporal, and band/layer parameters."

Rather inconveniently, it requires you to be signed in to NASA EarthData. Fortunately, it's quite easy (and, most importantly, free) to register an account, and the RemoteSensing paclet provides tools to log you in programmatically.

AρρEEARS allows you to access the real recorded data values for any available remote sensing product it offers. This makes it the appropriate tool to use if you are in need of accurate data for analysis or other scientific purposes. However, it is not so good for data exploration, as it can be fairly slow depending on the scale of the task and the availability of the API, and it will often return large numbers of heavy files. Use A$\rho\rho$EEARS if you already know what you want and you need precise and accurate data.

So let's see what A$\rho\rho$EEARS can do!

AppEEARSData: Product and Layer Metadata from A$\rho\rho$EEARS

Use the AppEEARSData function to get information about the available A$\rho\rho$EEARS geographic products and their layers.

List available A$\rho\rho$EEARS products:

In[ ]:= **AppEEARSData[] // Take[#, 10] &**

Out[ ]= {GPW_DataQualityInd, GPW_UN_Adj_PopCount,
    GPW_UN_Adj_PopDensity, GPW_Basic_Demog_Char, MCD12Q1,
    MCD12Q2, MCD15A2H, MCD15A3H, MCD43A1, MCD43A2}

Get the properties of an A$\rho\rho$EEARS product:

In[ ]:= **AppEEARSData["MOD17A2HGF", All] // Dataset**
    (*AppEEARSData["MOD17A2HGF","Dataset"]*)

Out[ ]=

Layers	{Gpp_500m, PsnNet_500m, Psn_QC_500m}
Platform	Terra MODIS
Description	Gross Primary Productivity (GPP)
RasterType	Tile
Resolution	500m
TemporalGranularity	8 day
Version	061
Available	True
DocLink	https://doi.org/10.5067/MODIS/MOD17A2HGF.061
Source	LP DAAC
TemporalExtentStart	2000–01–01
TemporalExtentEnd	Present
Deleted	False
DOI	10.5067/MODIS/MOD17A2HGF.061
Info	
ProductAndVersion	MOD17A2HGF.061

Get the value of a particular AppEEARS product property:

In[·]:= **AppEEARSData["MOD17A2HGF", "Resolution"]**

Out[·]= 500m

Get a list of available layers for a product:

In[·]:= **AppEEARSData["MOD17A2HGF"]**

Out[·]= {{MOD17A2HGF, Gpp_500m},
     {MOD17A2HGF, PsnNet_500m}, {MOD17A2HGF, Psn_QC_500m}}

Get the properties of an AppEEARS layer:

In[·]:= **AppEEARSData[{"MOD17A2HGF", "Gpp_500m"}]**
     (*AppEEARSData[{"MOD17A2HGF","Gpp_500m"},All]*)
     (*AppEEARSData[{"MOD17A2HGF","Gpp_500m"},"Dataset"]*)

Out[·]= <| AddOffset → 0., Available → True, DataType → float32, Description →
       MODIS/Terra Gross Primary Production (GPP) 8–Day L4 Global 500m SIN Grid,
       Dimensions → {time, YDim, XDim}, FillValue → 32 767,
       FillValueAll → {32 761, 32 762, 32 763, 32 764, 32 765, 32 766, 32 767},
       Group → , IsQA → False, Layer → Gpp_500m, OrigDataType → int16,
       OrigValidMax → 30 000, OrigValidMin → 0, QualityLayers → ['Psn_QC_500m'],
       QualityProductAndVersion → MOD17A2HGF.061, ScaleFactor → 0.0001,
       Units → kgC/m^2/8day, ValidMax → 3., ValidMin → 0., XSize → 2400, YSize → 2400 |>

Get the value of a particular AppEEARS layer property:

In[·]:= **AppEEARSData[{"MOD17A2HGF", "Gpp_500m"}, "Description"]**

Out[·]= MODIS/Terra Gross Primary Production (GPP) 8–Day L4 Global 500m SIN Grid

Get information about available projections for AppEEARS products:

In[·]:= **AppEEARSData["Projections"] // Dataset // Transpose**

Out[·]=

Name	{ ⋯10 }
Description	{ ⋯10 }
Platforms	{ ⋯10 }
Proj4	{ ⋯10 }
Datum	{ ⋯10 }
EPSG	{ ⋯10 }
Units	{ ⋯10 }
GridMapping	{ ⋯10 }
Available	{ ⋯10 }

AppEEARSAuthenticate: "Logging In"

If you like, you can manually authenticate with the A$\rho\rho$EEARS API. If you do not do this, you will be prompted by an authentication popup when you run a function requiring authentication in a kernel session for the first time:

In[·]:= **(*AppEEARSAuthenticate["Username","Password"]*)**

Out[·]= uZEFYMSYDOi6hXszD5umi7ppGLxCzLXNtoDwU6kuzFOPIV4J6TGvS8ebyzW–
VkzNC04boS0Uz3mbMlNb5qYsmw

MakeAppEEARSTaskRequest: Manually Opening API Tasks

If you like to tinker and keep your kernel free while your API task runs in the background, you can use MakeAppEEARSTaskRequest to start a new A$\rho\rho$EEARS task, specifying a {product, layer} pair and the request settings.

The function requires that you specify a parameter "BoundingBox" or "Points" in your request. The former can be any geographic region or GeoBoundingBox, while the latter must be a list of {lat, lon} pairs.

Here's how you could manually make an A$\rho\rho$EEARS task request.

An area request:

In[·]:= **task = MakeAppEEARSTaskRequest[{"MCD15A2H", "Lai_500m"},**
<| "BoundingBox" → Brittany, France ADMINISTRATIVE DIVISION **,**
"StartDate" → Mon 1 May 2023 **, "EndDate" → Today |>]**

Out[·]= AppEEARSTask[Task ID: 101756e7 –df70 –4c86 –bd47 –c054042af9a2]

A point request (supported, but not prioritized):

In[·]:= **task = MakeAppEEARSTaskRequest[{"VNP43IA2", "BRDF_Albedo_Uncertainty"},**
<| "Points" → {{0, 0}, {1, 1}, {−1, 1}},
"StartDate" → Mon 1 May 2023 **, "EndDate" → Today |>]**

Out[·]= AppEEARSTask[Task ID: e148a050 –a2e4 –4ffb –a308 –af98f8aed2e8]

MakeAppEEARSTaskRequest returns an AppEEARSTask object. You can use it to access information about the task and, once it completes, the files corresponding to your request.

AppEEARSTask: Asynchronously Calling Task Methods

The RemoteSensing paclet uses AppEEARSTask objects to represent A*ρρ*EEARS tasks. Here's a peek at what you can do with them.

Represent an A*ρρ*EEARS task given its task ID:

In[⋅]:= **task = AppEEARSTask["101756e7–df70–4c86–bd47–c054042af9a2"]**

Out[⋅]= AppEEARSTask[Task ID: **101756e7 –df70 –4c86 –bd47 –c054042af9a2**]

Request the task ID from the task object:

In[⋅]:= **task["TaskID"]**

Out[⋅]= 101756e7–df70–4c86–bd47–c054042af9a2

Request details about a task:

In[⋅]:= **task["Details"] // Dataset**

Out[⋅]=

tier	2
error	`Null`
params	`<| …₄ |>`
status	done
crashed	False
created	2024–02–06T21:11:11.665640
task_id	101756e7–df70–4c86–bd47–c054042af9a2
updated	2024–02–06T21:14:17.055016
user_id	phileasdg@gmail.com
attempts	1
estimate	$\{request\_size \to 3.64731 \times 10^{7}, request\_memory \to 1\}$
retry_at	`Null`
completed	2024–02–06T21:14:16.841195
has_swath	False
task_name	mcd15a2h–lai_500m–tue–6–feb–2024–16–11–07
task_type	area
api_version	v1
svc_version	3.46
web_version	`Null`
has_nsidc_daac	False

↗ ∧ rows 1–20 of **21** ∨ ↘

Request task status information:

In[ ]:= **task["StatusReport"] // Dataset**

Out[ ]=

tier	2
error	Null
params	{ ...4 }
status	done
crashed	False
created	2024–02–06T21:11:11.665640
task_id	101756e7–df70–4c86–bd47–c054042af9a2
updated	2024–02–06T21:14:17.055016
user_id	phileasdg@gmail.com
attempts	1
estimate	$\{\langle\,\vert\,\text{request\_size} \to 3.64731 \times 10^7\,\vert\,\rangle, \langle\,\vert\,\text{request\_}$
retry_at	Null
completed	2024–02–06T21:14:16.841195
has_swath	False
task_name	mcd15a2h–lai_500m–tue–6–feb–2024–16–11–0
task_type	area
api_version	v1
svc_version	3.46
web_version	Null
has_nsidc_daac	False

rows 1–20 of 21

Request a progress summary for a task:

In[ ]:= **task["ProgressSummary"] // Dataset**

Out[ ]=

status	done

Request the progress percentage:

In[ ]:= **task["ProgressPercent"]**

Out[ ]=   100%

List raw available file details from A*pp*EEARS for a completed task:

In[ ]:= **task["BundleDetails"] // Dataset**

Out[ ]=

files	{ ...111 }
created	2024-02-06T21:13:18.872866
task_id	101756e7-df70-4c86-bd47-c054042af9a2
updated	2024-02-06T21:13:59.834994
bundle_type	area

List available file details for a completed task (excluding supporting files):

In[ ]:= **task["SortedBundleDetails"] //**

      Take[#, 2] & // Dataset // Transpose // Dataset[#, ItemSize → 20] &

Out[ ]=

	MCD15A2H.061_2023114_to_2024	MCD15A2H.061_2023114_to_2024
sha256	e8564c0c4758df5d658d97cceb86c	265c510b3099e70310354ed3d33b
file_id	3f0aaa16-814a-4deb-9bbf-db01!	9fc0768b-4850-4a3e-9c7b-030b.
file_size	176 703	185 548
file_type	tif	tif
s3_url	s3://appeears-output/101756e7-c	s3://appeears-output/101756e7-c

List the names of available files grouped by date and layer:

In[ ]:= **task["SortedFileNames"] // Take[#, 2] & //**

      Take[#, 2] & // Dataset[#, ItemSize → 20] &

Out[ ]=

Tue 2 May 2023	MCD15A2H.061	Lai_500m	MCD15A2H.061_2023114_to_2024
	MCD15A2H.061	FparLai_QC	MCD15A2H.061_2023114_to_2024
	MCD15A2H.061	FparExtra_QC	MCD15A2H.061_2023114_to_2024
Wed 10 May 2023	MCD15A2H.061	Lai_500m	MCD15A2H.061_2023114_to_2024
	MCD15A2H.061	FparLai_QC	MCD15A2H.061_2023114_to_2024
	MCD15A2H.061	FparExtra_QC	MCD15A2H.061_2023114_to_2024

Download all available files from the completed request (this can take a long time):

In[ ]:= **(*task["GetFiles"]*)**

Download available files by their file names:

In[·]:= **ArrayPlot[♯, ColorFunction → Hue, ColorFunctionScaling → False] & /@ ImageData /@**
task["GetFiles", Take[Values[task["SortedFileNames"][[All, 1]]], 3]]

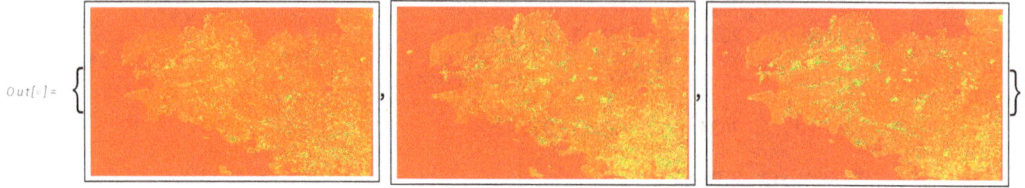

Out[·]=

A*ρρ*EEARSImages (for Small Requests)

If you are confident your task will resolve in a timely manner (i.e. you have a good internet
connection, the task is short and the A*ρρ*EEARS API is working normally*), you can use
AppEEARSImages to make a request, keep track of its progress and automatically download
the resulting data. Here's what that looks like.

Start a new task, monitor its progress automatically and automatically import desired data:

In[·]:= **AppEEARSImages[{"MCD15A2H", "Lai_500m"}, <|**

        "BoundingBox" → [Brittany, France ADMINISTRATIVE DIVISION] **,**

        "StartDate" → [Mon 1 May 2023] **, "EndDate" → Today |>] //**

        Take[♯, 3] & // Dataset

Out[·]=

Tue 2 May 2023	MCD15A2H.061	Lai_500m	
	MCD15A2H.061	FparLai_QC	
	MCD15A2H.061	FparExtra_QC	
Wed 10 May 2023	MCD15A2H.061	Lai_500m	
	MCD15A2H.061	FparLai_QC	
	MCD15A2H.061	FparExtra_QC	
Thu 18 May 2023	MCD15A2H.061	Lai_500m	
	MCD15A2H.061	FparLai_QC	
	MCD15A2H.061	FparExtra_QC	

*The A*ρρ*EEARS API has regular maintenance on Wednesdays, which may disrupt the service.

Acknowledgments

I would like to express my sincere gratitude to Christopher Wolfram for his support and guidance throughout this project. On many occasions, Christopher's advice and debugging skills were crucial in moving my work on this project forward. Without his help, I am sure it would not have moved as fast or gone as far as it has. My thanks extend also to Stephen Wolfram for suggesting this project to me, and for his kind words of support and enthusiasm at the sight of my nearly finalized project.

References

1. NASA, "Welcome to AρρEEARS!," *AρρEEARS*, accessed February 1, 2024. appeears.earthdatacloud.nasa.gov.

2. NASA, "Introduction," *Global Imagery Browse Services (GIBS)*, accessed February 1, 2024. nasa-gibs.github.io/gibs-api-docs.

Cite This Notebook

"RemoteSensing Paclet: GIBS Satellite Imagery and AρρEEARS Geographic Data Products"
by Phileas Dazeley-Gaist
Wolfram Community, STAFF PICKS, July 12, 2023
community.wolfram.com/groups/-/m/t/2959942

Efficient Discovery of Halting Paths in Aggregation System Multiway Graphs

PIETRO PEPE

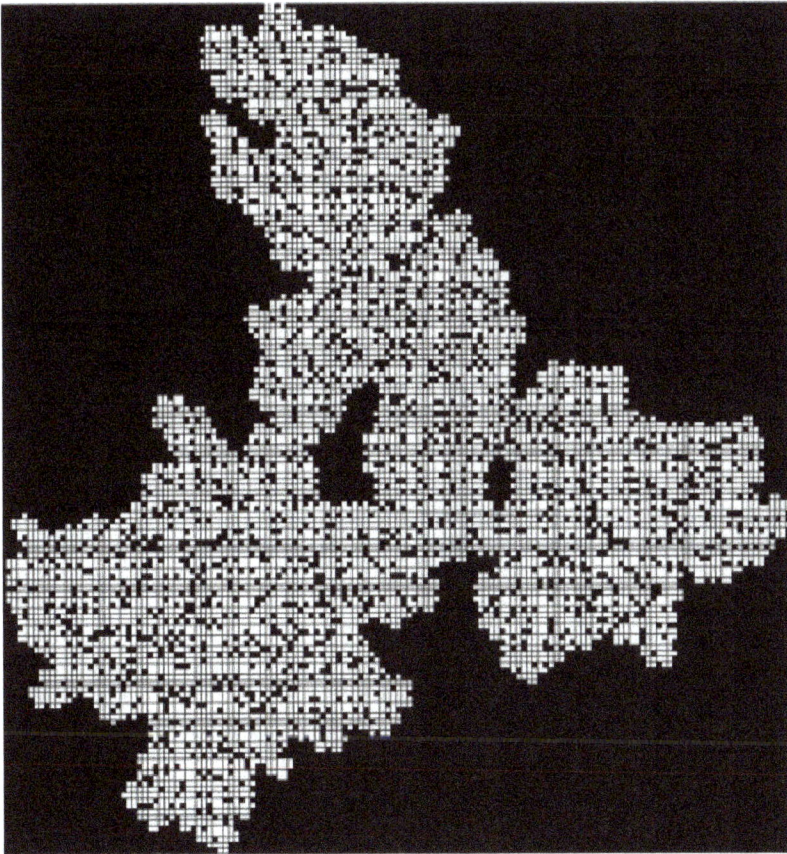

This project investigates totalistic aggregation systems, which exhibit cellular automata behavior with a unique twist. The focus is on totalistic rules that determine cell eligibility based on the total number of active neighboring cells. By developing a simulation software using Lua and the LÖVE framework, the project enables the exploration of different rule sets and initial conditions. Multiway graphs are used to visualize the system's behavior, and the concepts of translation, rotation and reflection canonicalization simplify the analysis of equivalent states. The project also proposes a new classification system—k-bit rules—that systematically identifies halting states based on rules and minimal initial conditions, and uncovers insights into system growth and eventual halting. Overall, this project provides valuable insights into totalistic aggregation systems and their dynamics.

Access the Full Code

Scan or visit wolfr.am/WSS2023-Pepe.

Introduction

Aggregation systems are like cellular automata (CAs) with a twist: once cells are set, i.e. aggregated, they remain set. The deterministic nature of a standard CA is here replaced by a (standardly random) choice of what cell to add to the system, from a list of eligible cells according to the system rule.

We focus our study on *totalistic rules*, rules that indicate whether a cell is eligible based on the total number of neighbor cells active instead of specific configurations around it.

A neighborhood can be von Neumann (2^4 rules) or Moore (2^8 rules), even though Moore is our main focus, and all graphs and simulations will be based on it, unless explicitly mentioned otherwise.

Here is an example of a totalistic Moore aggregation system evolution with rule 18 (10010). This system expands to cells with two or five active neighbors (the positions of bits set in the decimal representation):

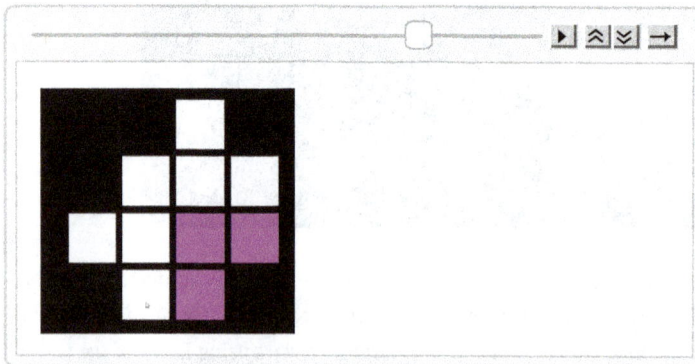

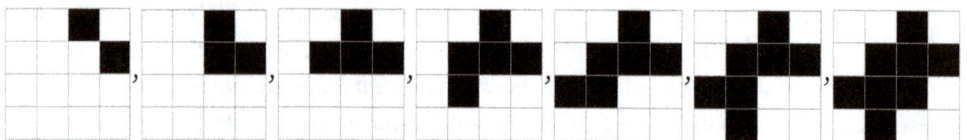

Depending on the cells selected, we can get to a state where no new cell can be added because none of them have an accepted quantity of active neighbor cells. We call this *halting,* and this is exactly what happens in the previous example when we reach .

Since we have potentially many ways to expand the system, it is natural to ponder how and when the aggregation choices affect the future of the system:

- When is it possible to make the system halt? When can it grow "forever"?
- When does a system always halt, regardless of path/choices? When does it never halt?
- "Haltable" systems always halt—that is, is there a halting path from any given state?

Our goal is to explore visualization and computation of halting states when starting from minimal initial states.

Simulation Software

We implement simulation software to explore any set of rules and initial conditions, with both manual and random selection of aggregated cells.

The simulation was developed with the framework LÖVE, using the Lua programming language. Executable files and source code are available at github.com/LexLoki /WSS23_AggSystems. **Note:** This notebook has a desktop version that has an integration with the simulator: clicking a node in a graph launches this external simulator with that specific node state set. For it to work, the executable folder of the simulator needs to be in the same directory as the notebook file.

In this software, aggregated cells are white and eligible cells are pink. Several halting states were found manually with the aid of the software:

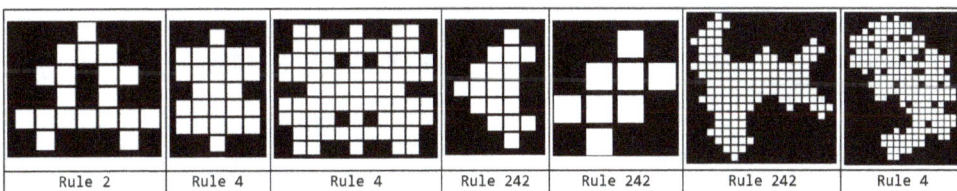

| Rule 2 | Rule 4 | Rule 4 | Rule 242 | Rule 242 | Rule 242 | Rule 4 |

In addition, a big, fancy one from rule 4 was found:

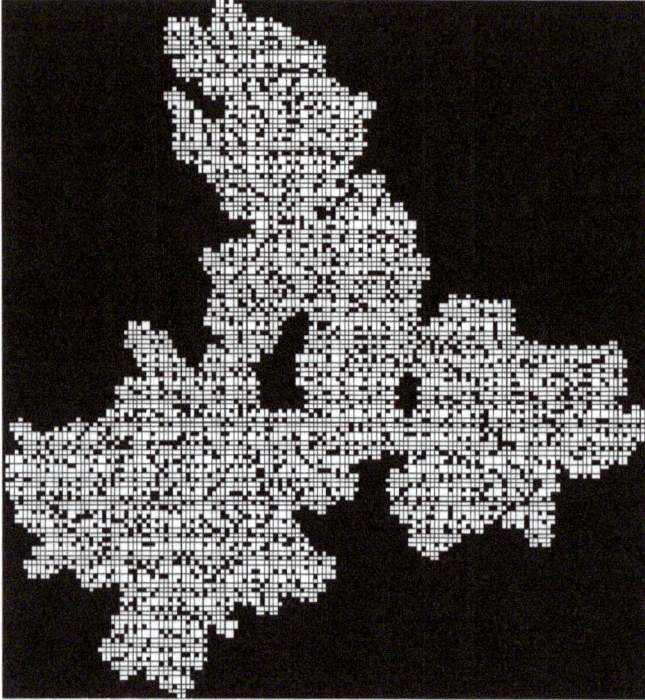

However interesting, these simulations are done following a single path at a time, and are thus far from a fully broad approach to understanding and visualizing these aggregation systems. So let's get started with using Mathematica.

Multiway Graphs

Since many choices can be made, it is natural to consider the multiway graph representation of these systems. These graphs represent all possible states a system can assume as vertices as well as viable transitions between two states as edges.

For instance, if we consider rule 255, which accepts all cells, starting with a single cell :

- One step:

In[•]:= **TotalisticGraph[1, 255, {{0, 0}}, True, True, "SpringElectricalEmbedding"]**

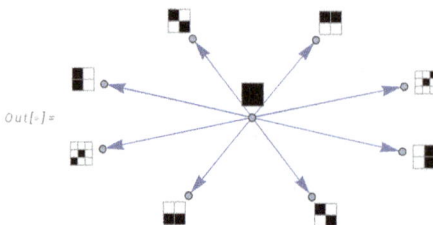

Out[•]=

- Two steps:

In[ ]:= **TotalisticGraph[2, 255, {{0, 0}}, True, True, "SpringElectricalEmbedding"]**

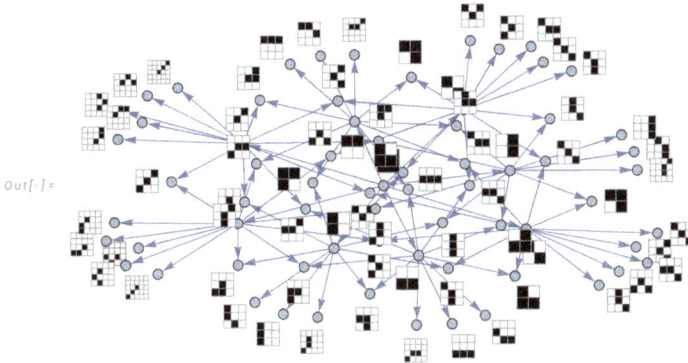

Out[ ]=

This shows that these systems can grow pretty fast, although this varies highly based on the rule. See, for instance, the multiway graphs of rule 2, which accept only cells with two neighbors, on , for 1, 2 and 3 steps:

In[ ]:= **Row[TotalisticGraph[#, 2, {{0, 0}, {1, 1}}, True, True, "SpringElectricalEmbedding"] & /@ Range[3], Spacer[10]]**

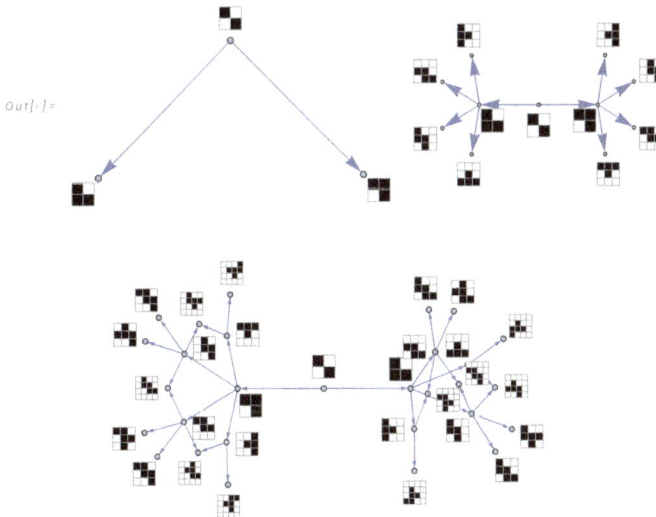

Out[ ]=

Unfortunately, rendering the graph takes a substantial amount of time, especially when generating the cells' plots for each vertex.

Rule 2 execution times demonstrate the following:

In[ ]:= **ParallelTable[**
    TotalisticGraph[n, 2, {{0, 0}, {1, 1}}, False, True] // VertexCount, {n, Range@12}]

Out[ ]= {3, 11, 29, 69, 157, 369, 861, 2017, 4867, 12 039, 30 329, 77 021}

```
In[•]:= (*steps = Range@12;
       MeasureTimes[plotCells_:False,moore_:True] := ParallelTable[
            TotalisticGraph[n,2,{{0,0},{1,1}},plotCells,moore]//Timing//First,{n,steps}]
          graphTimes = MeasureTimes[False,True];
       graphPlotTimes = MeasureTimes[True,True];
       Grid[Transpose[{Prepend[steps,"Steps"],Prepend[graphTimes,"Graph(seconds)"],
               Prepend[graphPlotTimes,"Graph+cells(seconds)"]}]//Transpose,Frame→All]*)
```

Steps	1	2	3	4	5	6	7	8	9	10	11	12
Graph(seconds)	0.	0.	0.	0.015625	0.015625	0.078125	0.203125	0.34375	1.20313	3.28125	7.625	20.7813
Graph+cells(seconds)	0.	0.0625	0.0625	0.15625	0.3125	0.890625	2.01563	4.89063	10.5625	25.0156	59.3594	139.641

In another scenario, rule 4 only accepts cells with three neighbors, starting with :

```
In[•]:= TotalisticManipulate[6, 4, {{0, 0}, {1, 1}, {0, 2}}, True, True]
```

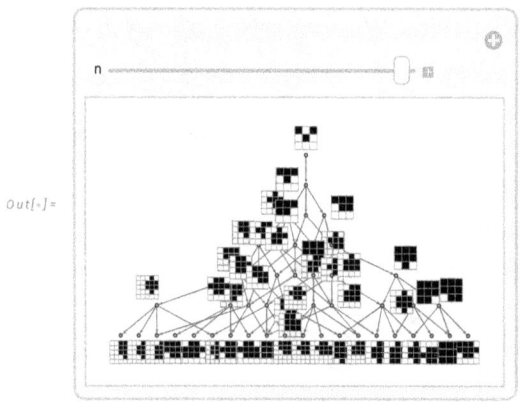

This multiway graph seems to be growing even slower than the previous one (rule 2). It is not hard to see why: this requires that more neighbors be active, so it is natural that fewer cells are eligible to be added to the system.

Even though the growth is lower, computations show it also "explodes," making it hard to study many interactions, especially when we analyze other rules:

- So far, we are not able to find any halting state, represented by a leaf in the multiway graph, and it is hard to visually inspect as we go further.

- One should note that these graphs have many states that are equivalent under translations, rotations and reflections, and so could be collapsed/merged in single states.

Introducing Translation Canonicalization

We want to apply some transformation on the set of cells that represent a state in a way that two different sets that are equivalent under translation, i.e. by an offset, map to the same value. One simple way to do this is to translate our state to the origin, or, more specifically, push the minimum boundary of the grid to the origin (top-left corner). This is what CellCanonicalize does. This allows more merges in the multiway graphs. For instance:

```
In[ ]:= ♯[3, 242, {{0, 0}, {1, −1}}, True] &/@ {TotalisticGraph, TotalisticCanonGraph} //
          Grid[Transpose[{♯, {"Rule 242 raw", "Rule 242 with canon translation"}}] // Transpose,
            Frame → All] &
```

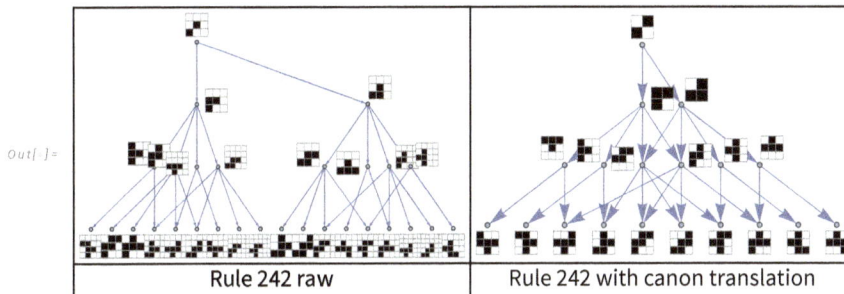

| Rule 242 raw | Rule 242 with canon translation |

We can see how that allows us to go deeper in reading the graph:

```
In[ ]:= ♯[4, 242, {{0, 0}, {1, −1}}, True, True, Automatic] &/@
          {TotalisticGraph, TotalisticCanonGraph} //
          Grid[Transpose[{♯, {"Rule 242 raw", "Rule 242 with canon translation"}}] // Transpose,
            Frame → All] &
```

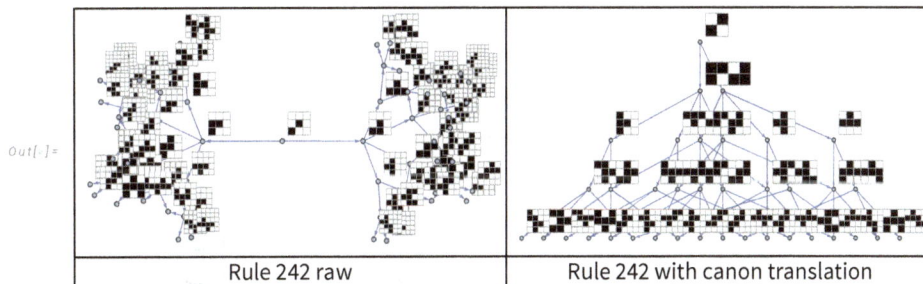

| Rule 242 raw | Rule 242 with canon translation |

The cost of performing the canonicalization is proportional to the quantity of cells of a given state. This cost is overly compensated by the amount of merges, thus reduced branching.

Rule 2 execution times demonstrate the following:

```
In[ ]:= steps = Range@12;
       MeasureTimes[plotCells_:False, moore_:True] := ParallelTable[
           TotalisticCanonGraph[n, 2, {{0, 0}, {1, 1}}, plotCells, moore] // Timing // First,
           {n, steps}]
       graphTimes = MeasureTimes[False, True];
       graphPlotTimes = MeasureTimes[True, True];
       Grid[Transpose[{Prepend[steps, "Steps"], Prepend[graphTimes, "Graph(seconds)"],
           Prepend[graphPlotTimes, "Graph+cells(seconds)"]}] // Transpose, Frame → All]
```

Steps	1	2	3	4	5	6	7	8	9	10	11	12
Graph (seconds)	0.	0.	0.	0.	0.	0.046875	0.0625	0.109375	0.28125	0.890625	2.28125	6.70313
Graph+cells (seconds)	0.	0.03125	0.046875	0.078125	0.21875	0.3125	0.765625	1.54688	3.67188	8.51563	21.7969	54.6719

But we can do better, since we still achieved considerable degrees of symmetry.

Introducing Rotational and Reflectional Canonicalization

The same idea applies here. The approach is to calculate for a given state all of its eight combinations of orthogonal rotation and reflection and map it to the first one, given some ordering algorithm. This is achievable in Mathematica using the resource function Array`. Rotations and relying on Order, through Sort. This is what CellCanonicalizeRotation does (it also does translation).

This greatly reduces the size of the graph in most cases, especially with symmetric and small initial conditions. It is easy to see why:

```
In[ ]:= ♯[3, 242, {{0, 0}, {1, −1}}, True, True, Automatic] &/@
           {TotalisticGraph, TotalisticCanonGraph, TotalisticFullCanonGraph} //
           Grid[Transpose[{♯, {"Rule 242 raw", "Rule 242 with canon translation",
               "Rule 242 full canon"}}] // Transpose, Frame → All] &
```

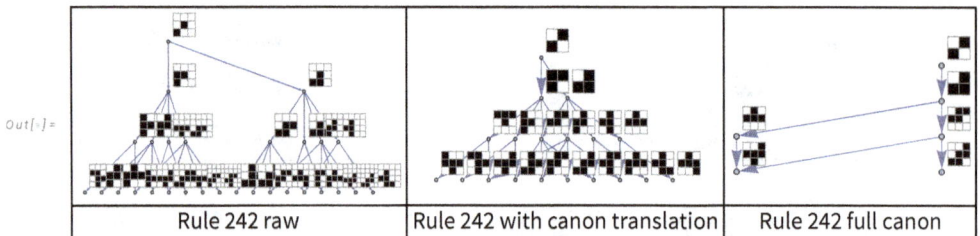

| Rule 242 raw | Rule 242 with canon translation | Rule 242 full canon |

One extra step comparing both canons:

```
In[·]:= #[5, 242, {{0, 0}, {1, −1}}, True, True, Automatic] &/@
         {TotalisticCanonGraph, TotalisticFullCanonGraph} //
         Grid[Transpose[{#, {"Rule 242 with canon translation", "Rule 242 full canon"}}] //
         Transpose, Frame → All] &
```

Out[·]=

| Rule 242 with canon translation | Rule 242 full canon |

Rule 2 execution times are even more impressive:

```
In[·]:= steps = Range@12;
       MeasureTimes[plotCells_:False, moore_:True] := ParallelTable[
           TotalisticFullCanonGraph[n, 2, {{0, 0}, {1, 1}}, plotCells, moore] // Timing // First,
           {n, steps}]
       graphTimes = MeasureTimes[False, True];
       graphPlotTimes = MeasureTimes[True, True];
       Grid[Transpose[{Prepend[steps, "Steps"], Prepend[graphTimes, "Graph(seconds)"],
           Prepend[graphPlotTimes, "Graph+cells(seconds)"]}] // Transpose, Frame → All]
```

Out[·]=

Steps	1	2	3	4	5	6	7	8	9	10	11	12
Graph (seconds)	0.	0.	0.	0.015625	0.015625	0.046875	0.046875	0.15625	0.15625	0.578125	1.57813	4.53125
Graph+cells (seconds)	0.	0.03125	0.03125	0.015625	0.03125	0.140625	0.046875	0.375	0.8125	1.5	4.20313	10.4844

This allows us to go even deeper.

Here are some other rules:

```
In[·]:= #[6, 4, {{0, 0}, {1, −1}, {2, 0}}, True, True, Automatic] &/@
           {TotalisticCanonGraph, TotalisticFullCanonGraph} //
         Grid[Transpose[{#, {"Rule 4 with canon translation", "Rule 4 full canon"}}] //
             Transpose, Frame → All] &
       #[5, 6, {{0, 0}, {1, −1}}, False, True, Automatic] &/@
           {TotalisticCanonGraph, TotalisticFullCanonGraph} //
         Grid[Transpose[{#, {"Rule 6 with canon translation", "Rule 6 full canon"}}] //
             Transpose, Frame → All] &
       #[5, 254, {{0, 0}, {1, −1}}, False, True, Automatic] &/@
           {TotalisticCanonGraph, TotalisticFullCanonGraph} //
         Grid[Transpose[{#, {"Rule 6 with canon translation", "Rule 6 full canon"}}] //
             Transpose, Frame → All] &
```

Out[◦]=

| Rule 4 with canon translation | Rule 4 full canon |

Out[◦]=

| Rule 6 with canon translation | Rule 6 full canon |

Out[◦]=

| Rule 6 with canon translation | Rule 6 full canon |

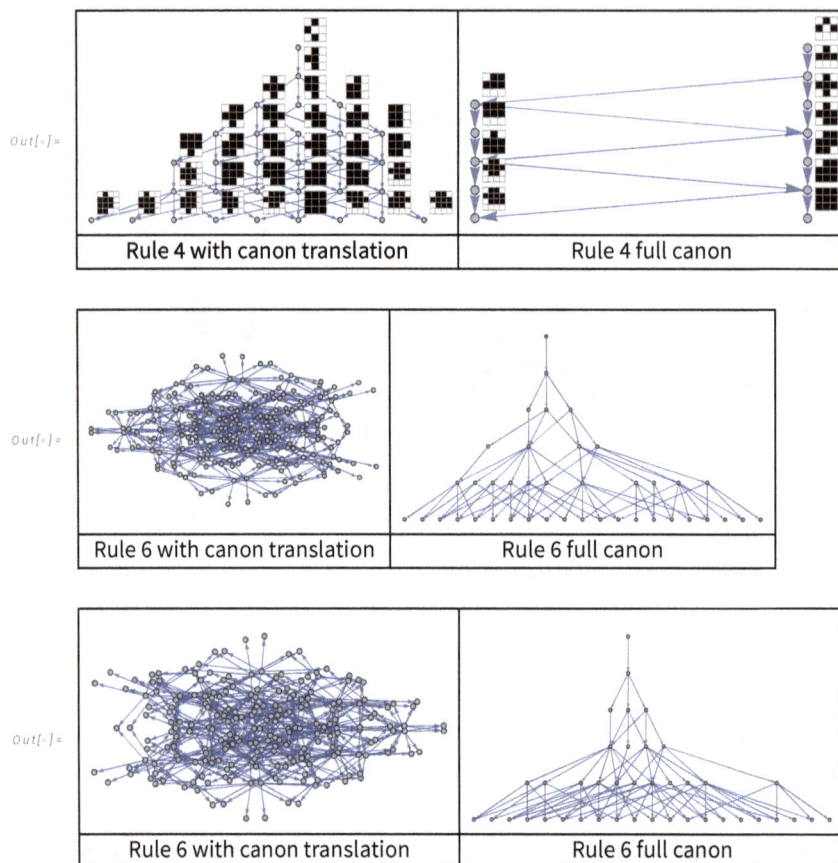

For visual exploration, this is great. With the added performance upgrade, we can probably try to start to look for halting states.

Halting States Computation

Halting states can be found by looking on the multiway system graphs, like in the one for rule 18:

In[◦]:= **TotalisticFullCanonGraph[7, 18, {{0, 0}, {1, −1}}, True, True] & // TimedEchoProcedure**

» Execution time 0.0625

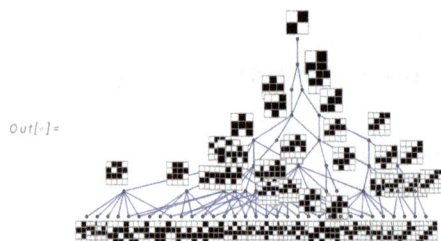

Out[◦]=

 is the only node in the tree that leads to nowhere. And that is right: all cells in the neighborhood of this state have a number of active neighbor cells different than 2, which is the only value accepted by totalistic rule 2.

Although this state might have been easy to find, depending on the rule and the initial state, they can be really hard (too deep) to find, sometimes even impossible. One challenge here is to be able to tell when we are not finding halting because it requires too many iterations, or because it indeed does not halt.

Moreover, an automated method for finding halting states is what we aim to create. It can be implemented by simply using these graphs. However, we can make it more efficient without the graph, since the graph, with or without the grid plots, also maps the edges. Since we are only interested in finding halts, we just need to keep the vertices; thus, we can replace NestGraph with an appropriate implementation using NestWhile and some modifications.

Trying to find the same halting rule 18 with our special method is quite simple—and, of course, runs faster:

```
In[ ]:= TimedHaltFindPlot[18, {{0, 0}, {1, −1}}, 7]
```

» Execution time 0.046875

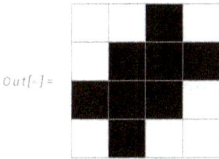

Out[ ]=

What about a harder-to-find halting rule? Rule 2 is one example where we cannot seem to find halting at depth 8:

```
In[ ]:= TotalisticFullCanonGraph[8, 2, {{0, 0}, {1, −1}}, True, True] & // TimedEchoProcedure
```

» Execution time 0.09375

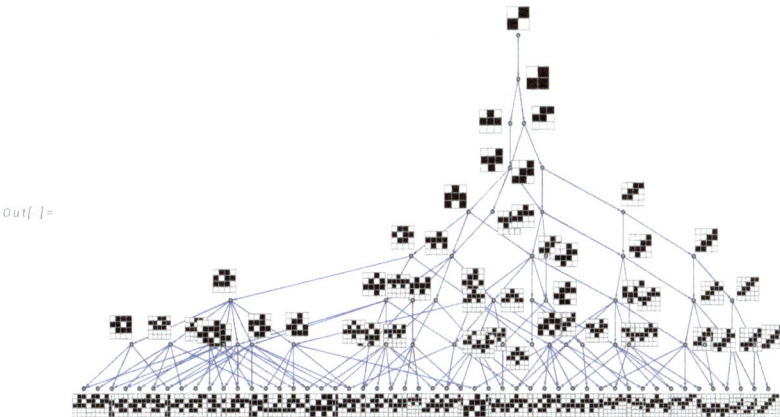

Out[ ]=

Trying more levels than that is unfeasible. We can find it, though:

In[ ]:= **TimedParallelHaltFindPlot[2, {{0, 0}, {1, −1}}, 11]**

» Execution time 0.25

Out[ ]=

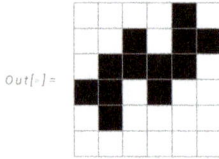

Now we are well-equipped to systematically study the set of possible rules, which will be done in both theory and practice (simulation) throughout the following sections.

Bit Groups and General Halting Proofs

As we know, the system rules are numbers whose binary digits represent the eligibility of a cell based on the quantity of its active neighbors. The less significant bit means one neighbor, while the most significant bit means eight neighbors. From here on out, when we say bit k, we refer to the bit of k-neighbor acceptance, or a shorthand for the acceptance.

We also adopt a distinct term, "k-bit rule," used to arrange our set of 255 possible rules (Moore/8 connectivity) in a relevant way for our analysis, as explained here.

Definition: A k-bit rule is a rule where the bit k is the lower-bit set, i.e. the system admits cells with exactly k neighbors and possibly more, but not fewer. B_k is the set of k-bit rules.

In[ ]:= **formattedData =**
  Transpose @ {Range @ Length @ bitGroups, StringJoin["(#", ToString @ Length @ #,
    "): ", StringRiffle[ToString /@ #, ","]] & /@ bitGroups};
 Print[Grid[Prepend[formattedData, {"Bit", "Bit Rules Set"}],
  Frame → All, Alignment → Left]];

Out[ ]=

Bit	Bit Rules Set
1	(#128): 1,3,5,7,9,11,13,15,17,19,21,23,25,27,29,31,33,35,37,39,41,43,45,47,49,51,53,55,57,59,61,63,65,67,69,71,73,75,77,79,81,83,85,87,89,91,93,95,97,99,101,103,105,107,109,111,113,115,117,119,121,123,125,127,129,131,133,135,137,139,141,143,145,147,149,151,153,155,157,159,161,163,165,167,169,171,173,175,177,179,181,183,185,187,189,191,193,195,197,199,201,203,205,207,209,211,213,215,217,219,221,223,225,227,229,231,233,235,237,239,241,243,245,247,249,251,253,255
2	(#64): 2,6,10,14,18,22,26,30,34,38,42,46,50,54,58,62,66,70,74,78,82,86,90,94,98,102,106,110,114,118,122,126,130,134,138,142,146,150,154,158,162,166,170,174,178,182,186,190,194,198,202,206,210,214,218,222,226,230,234,238,242,246,250,254
3	(#32): 4,12,20,28,36,44,52,60,68,76,84,92,100,108,116,124,132,140,148,156,164,172,180,188,196,204,212,220,228,236,244,252
4	(#16): 8,24,40,56,72,88,104,120,136,152,168,184,200,216,232,248
5	(#8): 16,48,80,112,144,176,208,240
6	(#4): 32,96,160,224
7	(#2): 64,192
8	(#1): 128

Remarks

- $R = \cup_{k=1}^{8} B_k$.

- If all bits set in a rule R_p are set in a rule R_q, the multiway system graph of R_p is contained in R_q. After all, a system with R_q can make all choices R_p can. We say R_q contains R_p.

    - See, for instance, how the multiway graph for rule 2 (binary 10) is contained in the one for rule 6 (binary 11):

In[ ]:= **TotalisticFullCanonGraph[3, ♯, {{0, 0}, {1, 1}}, True, True] &/@{2, 6} //**
    Row[♯, Spacer[8]] &

Out[ ]=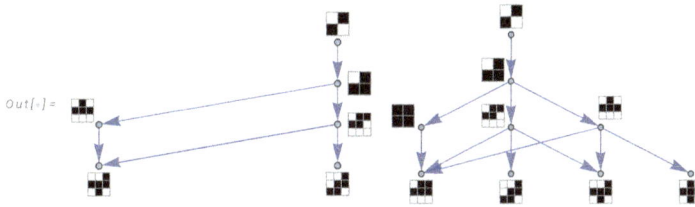

- The lower rule in a *k*-bit group is contained by all other rules in the same group. It should then be the easiest to compute.

- Analogously, the higher rule (in the same group) contains all others, and thus should be harder to compute.

4-Connectivity Halting Lemmas

You'll notice that 4-connectivity was not the focus of this study, and the following simple proofs make it clear why: they are far too simple in terms of determining halting. (Note that for 4-connectivity, there are only 16 rules and four bit groups.)

1-Bit Rules Never Halt

Proof: Let s be any given state (collection of cells). There will be a set of cells with higher horizontal coordinates. Let p be the coordinate of the cell from this set with higher vertical coordinates. The cells at positions p + {–1, 1} and p + {1, 1} will be unset, so the cell at p + {0, 1} will have exactly one neighbor. The system can grow with it:

Non-1-Bit Rules Always Halt

Proof: Let s be any given state (collection of cells). Let r be any rectangle that contains all cells from s. A cell outside r will have, at most, one neighbor cell inside rUs. It is then impossible to add a cell from outside, and so the system will eventually halt:

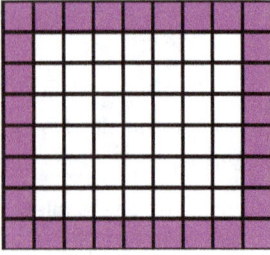

8-Connectivity Halting Lemmas

We can use similar arguments from 4-connectivity in 8-connectivity, but the conclusion results only apply to a small subset of the rules.

1-Bit Rules Never Halt

Proof: Let s be any given state (collection of cells). There will be a set of cells with higher horizontal coordinates. Let p be the coordinates of the cell from this set with higher vertical coordinates. The cell at position p + {1, 1} will have exactly one neighbor. The system can grow with it:

($k \geq 4$)-Bit Rules Always Halt

Proof: Let s be any given state (collection of cells). Let r be any rectangle that contains all cells from s. A cell outside r will have at most three neighbor cells inside rUs. It is then impossible to add a cell from outside, and so the system will eventually halt:

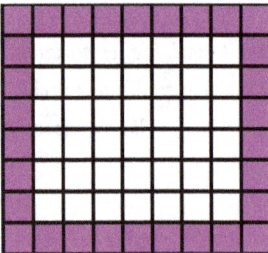

Now we are left with the set of 2-bit and 3-bit rules, 98 in total. We computationally analyze them in the following section. However, there is one interesting result.

2-Bit Rules with a Bit-3 Set Never Halt

Proof: We prove my contradiction by assuming we have a halting state—that is, a state where there are no eligible cells to insert. In this case, there is no cell with exactly two or three neighbors.

Assume that we have a halting state. We are going to define a "walk" along its boundary. We pick the cell at the rightmost of the top row of the boundary, like the gray cell here:

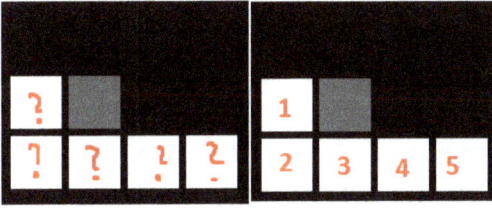

Apart from the gray corner cell, where we currently are, in general we do not know which cells are active. Breaking up the possibilities, we have the following:

- *If 4 is on and 5 is off, we have a double corner. The cell to our right has three neighbors, and so is eligible: contradiction.*

- *If 2 is on, the cell on our top has two neighbors, and so is eligible: contradiction.*

- *If 4 and 5 are off, and since every cell that was added required at least two neighbors, cells 2 and 3 are the only possible neighbors for the corner cell. So they are on, and our right has two neighbors, and so is eligible: contradiction.*

- *Otherwise, we move to the bottom row, go to the rightmost cell (our new corner) and repeat the previous checks.*

This process is finite, because otherwise we would be moving in a down-right direction infinitely, contradicting our hypothesis that we are walking on a finite boundary.

Conclusion: such a halting configuration is impossible.

This reduces our halting study set to 64 rules.

Bit Group Halt-Finding Analysis and Results

PlotBinaryGroupsHalt gives us the list of halting states found for a given rule, initial cells and maximum depth. A k-bit rule requires a minimum of k initial cells in order to not halt instantly (trivially). MinimalInitials provides the set of minimal initial cells for a given rule.

1-Bit

As we already know, we cannot halt with 1-bit rules:

In[ ]:= **PlotBinaryGroupsHalt[1, {{0, 0}}, 8]**

Out[ ]= { {}, {}, {}, {}, {}, {}, {}, {}, {}, {}, {}, {}, {}, {}, {}, {}, {}, {}, {},
    1   3   5   7   9 11 13 15 17 19 21 23 25 27 29 31 33 35 37 39

     {}, {}, {}, {}, {}, {}, {}, {}, {}, {}, {}, {}, {}, {}, {}, {}, {}, {}, {},
     41 43 45 47 49 51 53 55 57 59 61 63 65 67 69 71 73 75 77

$$\{\}, \{\}, \{\}, \{\}, \{\}, \{\}, \{\}, \{\}, \{\}, \{\}, \{\}, \{\}, \{\}, \{\}, \{\}, \{\},$$
79 81 83 85 87 89 91 93 95 97 99 101 103 105 107 109 111

$$\{\}, \{\}, \{\}, \{\}, \{\}, \{\}, \{\}, \{\}, \{\}, \{\}, \{\}, \{\}, \{\}, \{\}, \{\},$$
113 115 117 119 121 123 125 127 129 131 133 135 137 139 141

$$\{\}, \{\}, \{\}, \{\}, \{\}, \{\}, \{\}, \{\}, \{\}, \{\}, \{\}, \{\}, \{\}, \{\}, \{\},$$
143 145 147 149 151 153 155 157 159 161 163 165 167 169 171

$$\{\}, \{\}, \{\}, \{\}, \{\}, \{\}, \{\}, \{\}, \{\}, \{\}, \{\}, \{\}, \{\}, \{\},$$
173 175 177 179 181 183 185 187 189 191 193 195 197 199

$$\{\}, \{\}, \{\}, \{\}, \{\}, \{\}, \{\}, \{\}, \{\}, \{\}, \{\}, \{\}, \{\}, \{\},$$
201 203 205 207 209 211 213 215 217 219 221 223 225 227

$$\{\}, \{\}, \{\}, \{\}, \{\}, \{\}, \{\}, \{\}, \{\}, \{\}, \{\}, \{\}, \{\}, \{\}\}$$
229 231 233 235 237 239 241 243 245 247 249 251 253 255

2-Bit

In[·]:= **Echo [PlotMinimalInitials[2], "Initial states: "];**

» Initial states:

Initial Redundancy

It is easy to see that both initial states result in the same configurations, regardless of the subsequent bits:

In[·]:= **TotalisticGraph[1, 2, ♯, True] & /@ MinimalInitials[2] // Row[♯, Spacer[18]] &**

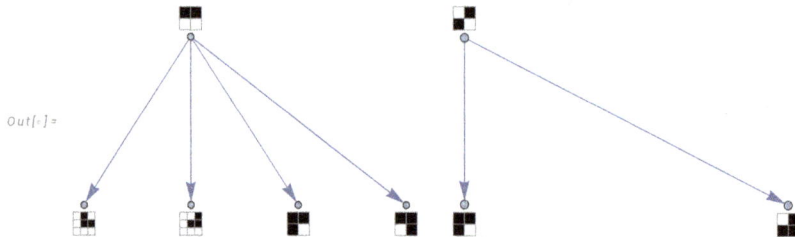

Out[·]=

Halting and Performance

For this rule, the branching factor can be substantially high. Computing only states (no edges) with HaltFind is both considerably lighter on memory and faster. Running 11 steps in the simplest rule gives us a halting configuration:

In[·]:= **TimedHaltFindPlot[First@bitGroups[[2]], First@MinimalInitials@2, 11]**

» Execution time 1.25

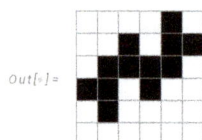

Out[·]=

This might not seem like much, but this is the simplest case for the 2-bit group, and thus represents a lower bound. Rules with more branching can get more expensive to run. Here is the most complicated rule (254) from this group, for instance:

In[ ]:= **TimedHaltFindPlot[Last@bitGroups〚2〛, First@MinimalInitials@2, 11]**

» Execution time 29.375

Out[ ]= No halt found

Our parallelized implementation can improve this, though:

In[ ]:= **TimedParallelHaltFindPlot[Last@bitGroups〚2〛, First@MinimalInitials@2, 11]**

» Execution time 4.85938

Out[ ]= No halt found

Although this improves the time to compute for a single rule, it does not work well for running for all the rules. Parallelizing across the sixty-four 2-bit rules, the sequential algorithm is significantly faster than sequentially running each parallelized algorithm. It is not possible in Mathematica to nest parallelizations, so we can only use one.

Let's see the halting states we can find, up to 11 steps, for each 2-bit rule. Our parallelized computation runs for a couple of minutes and results in the following:

In[ ]:= **(*PlotBinaryGroupsHalt[2,{{0,0},{1,1}},11, True])**

» Execution time 317.915

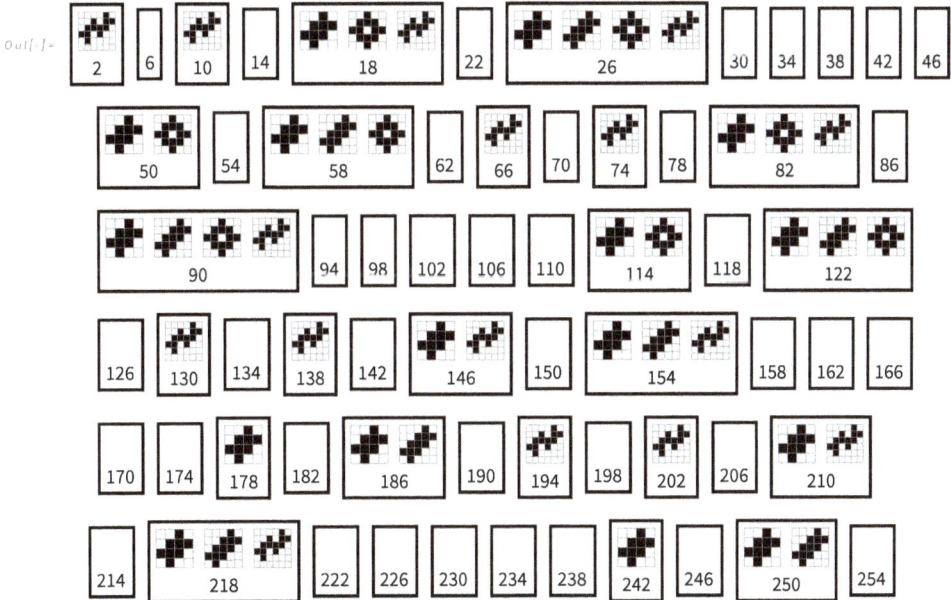

Remember that some of these rules (half of them) we already proved cannot halt:

```
In[ ]:=  FromDigits[Reverse@#, 2] &/@
            Select[Tuples[{0, 1}, 8], And[#[[1]] == 0, #[[2]] == 1, #[[3]] == 1] &] //
            Sort // Echo[#, "Provable never halt"] &;
```

» Provable never halt {6, 14, 22, 30, 38, 46, 54, 62, 70, 78, 86, 94, 102, 110,
    118, 126, 134, 142, 150, 158, 166, 174, 182, 190, 198, 206, 214, 222, 230, 238, 246, 254}

The rules we have not been able to determine whether they can converge are 34, 42, 98, 106, 162, 170, 226 and 234.

For the 2-bit rules, we will provide some interpretation/discussion of minimal cases that appeared.

Subcase Analysis:

Rule 2 is the lesser rule of the 2-bit group, so its halting state is a state reachable by every other rule in the group. Even though all other rules can reach this state, they do not necessarily get stuck on it. The condition to halt is easy to see when we observe the boundary of the system. The two neighbor cells in the "side holes" require bit 6 to expand, while all others require either bits 1 or 3. This means that having bits 3 and 6 unset makes this state a halting state. This means a quarter of the 64 cells in the 2-bit group reach this halting state:

```
In[ ]:=  FromDigits[Reverse@#, 2] &/@
            Select[Tuples[{0, 1}, 8], And[#[[1]] == 0, #[[2]] == 1, #[[3]] == 0, #[[6]] == 0] &] // Sort
```

Out[ ]= {2, 10, 18, 26, 66, 74, 82, 90, 130, 138, 146, 154, 194, 202, 210, 218}

Interestingly, for many of them, we can find a halting state with even fewer steps than rule 2: .

Subcase Analysis:

It is easy to see that a required condition for this state to halt is that the rule cannot have bit 3. This is all we need to get stuck in this state (since every cell has three or fewer neighbors), but we need a little bit more to be able to get in this state. Analyzing the tree, we can see how we got there:

```
In[ ]:=  Row[ShowSteps[{{1, 3}, {2, 4}, {2, 3}, {2, 2}, {3, 2}, {3, 1}, {4, 2}, {3, 3}}, 2], Spacer[1]]
```

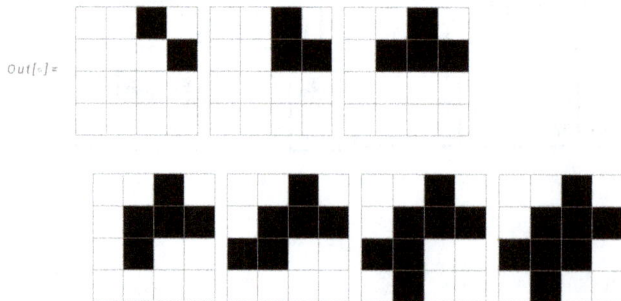

Out[ ]=

The last block required bit 5 to be set; all others required only bit 2. But what if we consider adding this last block earlier?

In[·]:= **Row[ShowSteps[{{1, 3}, {2, 4}, {2, 3}, {2, 2}, {3, 2}, {3, 3}, {3, 1}, {4, 2}}, 2], Spacer[1]]**

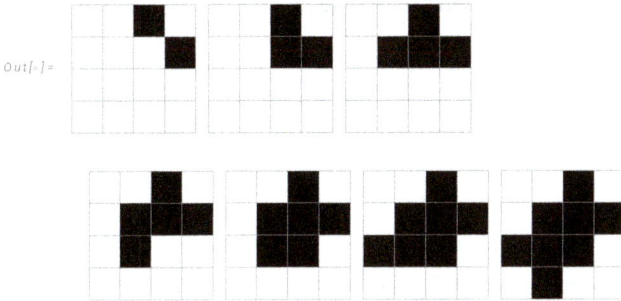

Out[·]=

The last block required bit 3 to be set, and that is a contradiction. That means it is impossible to follow this other order. **Conclusion:** This halting state happens when bit 3 is unset and bit 5 is set. These rules are:

In[·]:= **FromDigits[Reverse@#, 2] &/@**
Select[Tuples[{0, 1}, 8], And[#[[1]] == 0, #[[2]] == 1, #[[3]] == 0, #[[5]] == 1] &] // Sort

Out[·]= {18, 26, 50, 58, 82, 90, 114, 122, 146, 154, 178, 186, 210, 218, 242, 250}

Exactly the ones seen in the group plot halt as expected.

Special Subcase Analysis:

This state was found manually and requires 19 steps, using only bit-2 insertions. To halt, though, requires that bits 3, 4 and 6 be unset. Having bit 7 set lets the system grow two steps, adding the cells in the interior, and then also halts:

In[·]:= **FromDigits[Reverse@#, 2] &/@Select[Tuples[{0, 1}, 8],**
And[#[[1]] == 0, #[[2]] == 1, #[[3]] == 0, #[[4]] == 0, #[[6]] == 0] &] // Sort

Out[·]= {2, 18, 66, 82, 130, 146, 194, 210}

These rules are a subset of the rules that reach the lesser rule-halting state .

3-Bit

» Initial states:

» Reduced initial states:

(Reduction explained in the following section.)

Initial Redundancy

Initials ⬚ and ⬚ are equivalent after one step:

In[ ]:= **TotalisticGraph[1, 4, #, True] & /@ MinimalInitials[3]⟦{1, 3}⟧ // Row[#, Spacer @ 1] &**

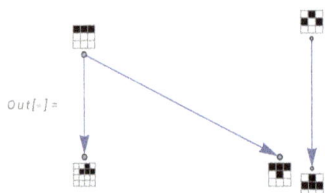

Out[ ]=

- Not only that, ⬚ is different after one step, but after two steps is also equivalent to ⬚, ⬚:

In[ ]:= **TotalisticFullCanonGraph[3, 4, #, True] & /@ MinimalInitials[3]⟦{3, 4}⟧ //**
Row[#, Spacer @ 1] &

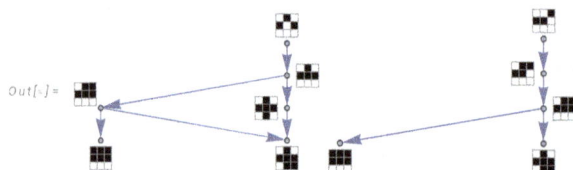

Out[ ]=

- Apart from that, ⬚ is quite trivial, since it halts after one step.

- ⬚ is a unique combination of three cells in a 3 × 3 grid, but is still invalid as it provides no eligible cell to be inserted.

Note that all these observations do not depend on the subsequent bits of the rule, since for these rules the first two steps have only 3-bit eligible cells.

Halting and Performance

The 3-bit rules have a substantially lower branching factor. We can compute several more steps than in 2-bit rules (around 10 more steps). First, we begin looking for halting with the lesser (simplest) 3-bit rule, rule 4, that we can find that halts in 15 steps:

In[ ]:= **TimedHaltFindPlot[First @ bitGroups⟦3⟧, First @ MinimalInitials @ 3, 15]**

» Execution time 0.53125

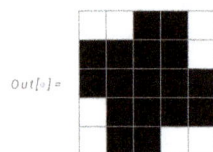

Out[ ]=

Running with the more complicated 3-bit rule, rule 252, we cannot find halting, even though we are able to run way more steps (21):

In[ ]:= **TimedHaltFindPlot[Last@bitGroups[[3]], First@MinimalInitials@3, 21]**

» Execution time 47.4688

Out[ ]= No halt found

Let's see the minimum halting states we can find, up to 11 steps, for each 3-bit rule:

■ Haltings for initial :

In[ ]:= **PlotBinaryGroupsHalt[3, MinimalInitials[3][[1]], 17, True]**

» Execution time 24.8788

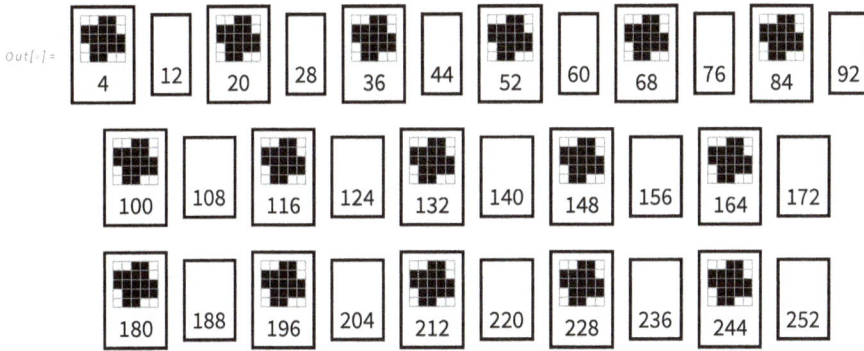

Out[ ]=

■ Haltings for initial :

In[ ]:= **PlotBinaryGroupsHalt[3, MinimalInitials[3][[5]], 17]**

» Execution time 99.0453

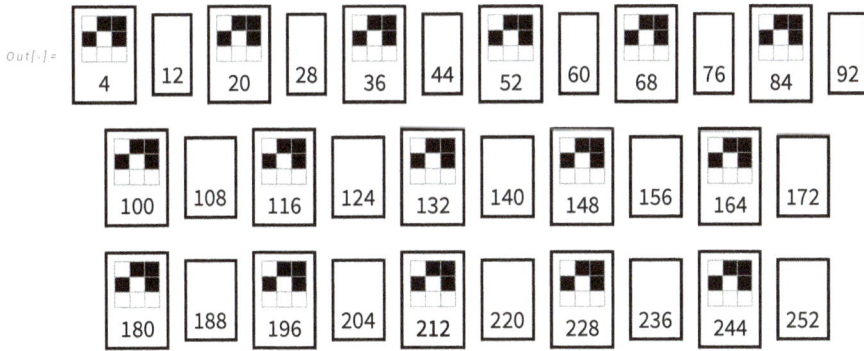

Out[ ]=

■ Haltings for initial ▨ :

In[ ]:= **PlotBinaryGroupsHalt[3, MinimalInitials[3][[6]], 17]**

» Execution time 149.177

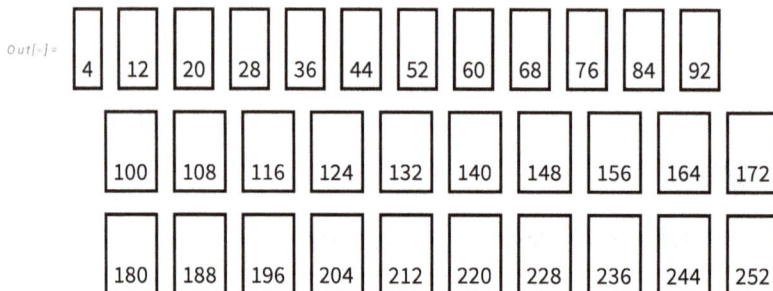

Out[ ]=

| 4 | 12 | 20 | 28 | 36 | 44 | 52 | 60 | 68 | 76 | 84 | 92 |

| 100 | 108 | 116 | 124 | 132 | 140 | 148 | 156 | 164 | 172 |

| 180 | 188 | 196 | 204 | 212 | 220 | 228 | 236 | 244 | 252 |

■ Haltings for initial ▨ :

In[ ]:= **PlotBinaryGroupsHalt[3, MinimalInitials[3][[7]], 17]**

» Execution time 238.189

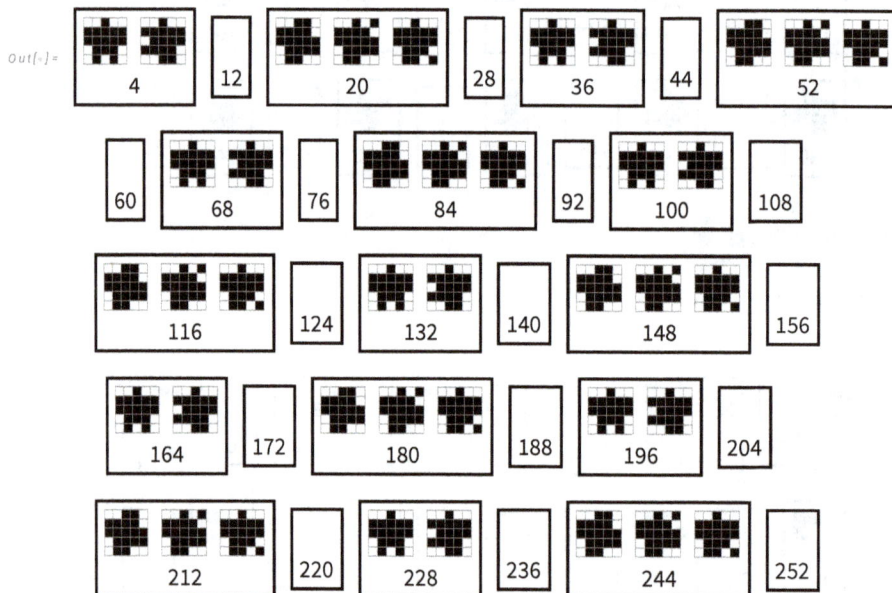

Out[ ]=

■ Haltings for initial :

In[ ]:= **PlotBinaryGroupsHalt[3, MinimalInitials[3][[9]], 17]**

» Execution time 150.67

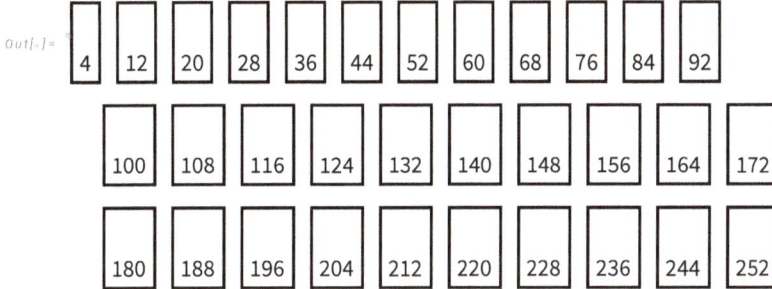

Out[ ]=

■ Haltings for initial :

In[ ]:= **PlotBinaryGroupsHalt[3, MinimalInitials[3][[10]], 17]**

» Execution time 255.397

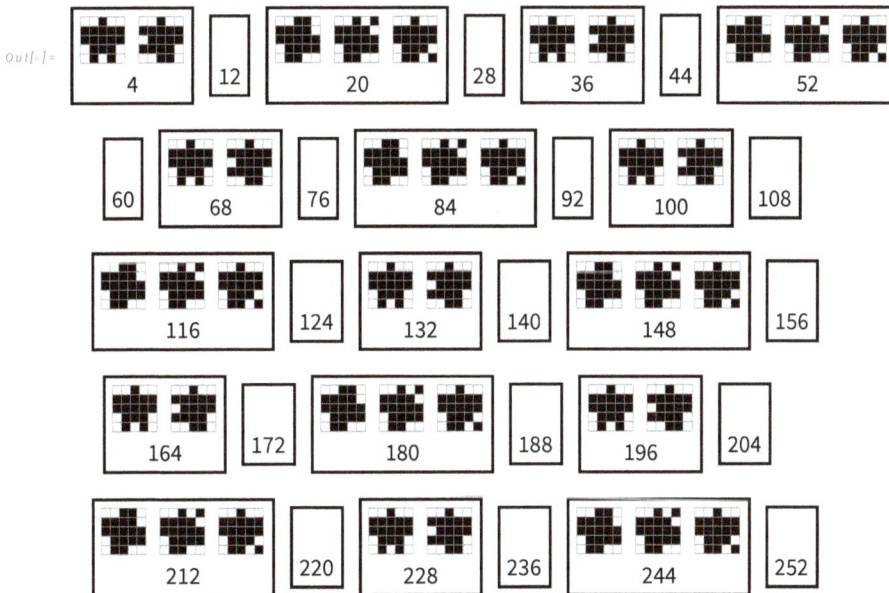

Out[ ]=

It is interesting that initial states have this peculiar effect in allowing halting states. Since we were not able to run these further, it is hard to say for sure if they made the halting possible or just made the halting possible sooner. The latter seems more likely, unless some pattern created by the initial case propagates infinitely in a way that prevents otherwise-halting configurations.

A look at one of these graphs shows an interesting pattern of steady branching until around a 3×3 grid and then starts to expand. See rule 4 (binary 100, the simplest of the group):

In[ ]:= **TotalisticFullCanonGraph[11, 4, {{0, 0}, {1, 1}, {0, 2}}, True]**

Out[ ]=

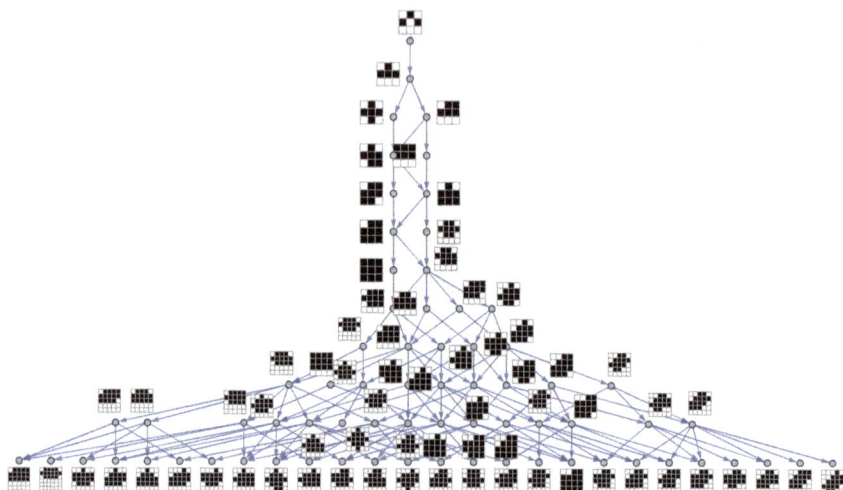

Concluding Remarks

In this project, we explored totalistic aggregation systems with minimal initial conditions, which proved to be a challenging yet intriguing subset of aggregation systems. Our efforts resulted in the development of a diverse collection of visual computations for these systems, offering valuable tools for further exploration and optimization. However, due to the combinatorial nature of the problems involved, significant optimizations beyond specific purposes or systems with particular properties are unlikely.

Nonetheless, we made significant contributions to the understanding of these systems. By proposing a relevant classification of rules, we focused on a subset of 96 rules and demonstrated that 32 of them never halt, further narrowing down the "gray area for conclusions" to a group of 64 rules. Among these, 24 rules (eight 2-bit and sixteen 3-bit) have yet to reveal a halting state, leaving the question open as to whether they can halt or not. The remaining 40 rules were shown to halt, but we have not determined which of their halting configurations represent possibilities in higher-branching steps of their multiway graphs.

Looking ahead, there are several promising avenues for future research:

- The sixteen 3-bit rules that resisted halting have a shared property that suggests they may be impossible to halt, similar to the proven non-halting nature of the 2-bit subset. Constructing a definitive proof for this observation would be a significant step forward.

- Further in-depth exploration of 2-bit and 3-bit rules, especially 3-bit sensitivity to some initial configurations, would also be a step in the right direction.

- We considered an alternative halt-finding method, involving pruning subgraphs of the multiway graph that are deemed unlikely to halt soon using greedy and heuristic criteria. Although not developed in this study, it presents an intriguing approach for future investigation.

- The simulator offers the capability to simulate random paths, opening up possibilities for individual "smart" path-finding queries to explore and discover halting states. Developing cell-selection policies for these paths could prove fruitful, providing an alternative to the classical broad-breadth search on multiway graphs.

Acknowledgments

I would like to thank all Wolfram staff involved in the 2023 Wolfram Summer School for giving me this opportunity to challenge myself in this exciting environment and assisting me along the way.

Special thanks to my mentor Robert Nachbar for all the assistance he provided me and other students, with his keen eye, technical knowledge and communication skills.

To Eric Parfitt, who helped me save (or avoid spending) countless hours with some technical issues. All with a glance of an eye.

And last but not least, to Stephen Wolfram, who introduced me to this project and made sure I would have a great project on hand, despite how picky I am.

References

1. S. Wolfram (2002), *A New Kind of Science*, Wolfram Media, Inc. www.wolframscience.com/nks.

2. K. Khanna (2019), "Aggregation Systems: A Stochastic Approach to CA," *Wolfram Community*. community.wolfram.com/groups/-/m/t/1728662.

3. L. M. Ruiz G. (2021), "Study Aggregation Systems with Multiway Graphs," *Wolfram Community*. community.wolfram.com/groups/-/m/t/2312514.

Cite This Notebook

"Efficient Discovery of Halting Paths in Aggregation System Multiway Graphs"
by Pietro Pepe
Wolfram Community, STAFF PICKS, July 12, 2023
community.wolfram.com/groups/-/m/t/2959726

Turing Machines on Graphs

STEFAN GRAHAM

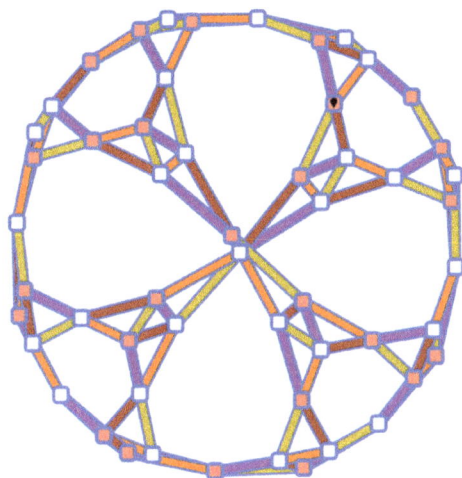

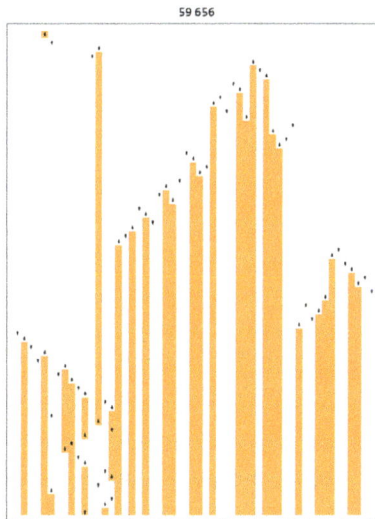

Ordinary Turing machines are implemented on infinite 1D tapes. This project investigates and explores the behavior of Turing machines on finite cyclic graphs. The graphs investigated are the circular cyclic graph, a torus and the Sierpiński network. This post demonstrates that the Sierpiński network exhibits the most complex and varied behavior.

Introduction to Turing Machines on Graphs

An ordinary Turing machine consists of a linear tape divided into cells, where each cell consists of a color and a state as indicated by the machine's head. The machine's head moves left and right along the tape, updating the state and color based on specified transition rules. In a graph-based Turing machine, the machine's computation takes place on a network of vertices and edges instead of on a linear tape. The Turing machine on a graph moves from node to node, following the edges based on the current state and the symbol it encounters. Whereas a traditional Turing machine only moves left and right, the Turing machine on a graph can move in any direction, only constrained by the number of adjacent edges.

In this post, the notation "(state, color)-TM" is used to refer to a specific Turing machine. For example, a 1-state, 2-color Turing machine is denoted as "(1, 2)-TM."

Rule Definition

The number of possible rules for a given Turing machine is determined by the number of states, colors and edges (if it is a graph). The total number of possible rules is given as:

(edges states colors)^(states colors)

For example, for a 1-state, 2-color linear tape (2-edges), there are 16 rules:

```
In[•]:= (2 × 1 × 2) ^ (1 × 2)
```

```
Out[•]= 16
```

For a 2-state, 2-color graph Turing machine with four edges, there are 65,536 rules:

```
In[•]:= (4 × 2 × 2) ^ (2 × 2)
```

```
Out[•]= 65 536
```

Given the number of states, colors and edges in a Turing machine, the rules governing a specific Turing machine are defined in the following function:

```
In[•]:= netTMRules[state_Integer, color_Integer, edges_Integer, ruleNr_Integer] :=
        Thread[Tuples[{Range[1, state], Range[0, color – 1]}] →
            Tuples[Tuples[{Range[1, state], Range[0, color – 1], Range[1, edges]}],
                {state * color}][[ruleNr]]]
```

The format for the rules is as follows:

{state, color} -> {state, color, next edge to follow}

For example, for 2-states, 2-colors and 2-edges, we have rule 441 specified as:

```
In[•]:= netTMRules[2, 2, 2, 441]
```

```
Out[•]= {{1, 0} → {1, 0, 1}, {1, 1} → {2, 1, 1}, {2, 0} → {2, 1, 2}, {2, 1} → {1, 0, 1}}
```

Example of a Turing Machine Evolution on a Graph

The simplest possible graph-based Turing machine is the cyclic graph. Essentially, this is a linear Turing machine where the two ends of the tape are glued together. A small-cycle Turing machine with four nodes is shown here to illustrate the analysis that will later be performed on more complex graphs:

```
In[•]:= {adjacencyList, edges, verticies} = cyclicGraph[4, 1];
        nrEdges = 2;
        inputGraph = adjacencyList;
```

In[·]:= **SimpleGraph[Graph[verticies, edges], VertexLabels → Automatic]**

Out[·]=

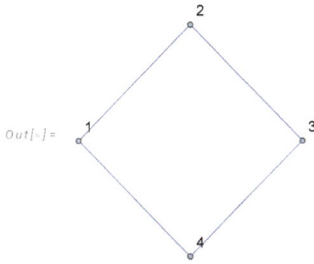

The Turing machine starts with a graph with given initial conditions indicating the color of each node and the position of the head. The machine under investigation starts with a colored node at vertex three and all other cells white:

In[·]:= **startVertex = 3;**
intialState =
{1, ReplacePart[Table[0, Length[inputGraph]], startVertex → 1], startVertex}

Out[·]= {1, {0, 0, 1, 0}, 3}

With a 1-state, 2-color Turing machine, rule 14 gives the following behavior:

In[·]:= **nrStates = 1; nrColors = 2; nrIterations = 16; nrRule = 14;**

In[·]:= **inputRule = NetTMRules[nrStates, nrColors, nrEdges, nrRule]**

Out[·]= {{1, 0} → {1, 1, 2}, {1, 1} → {1, 0, 2}}

In[·]:= **outEvolution = NetTMEvolveList[inputRule, inputGraph, intialState, nrIterations];**

In[·]:= **ListAnimate[**
Table[NetTMGraphics[SimpleGraph[Graph[verticies, edges]], l], {l, outEvolution}]]

Out[·]=

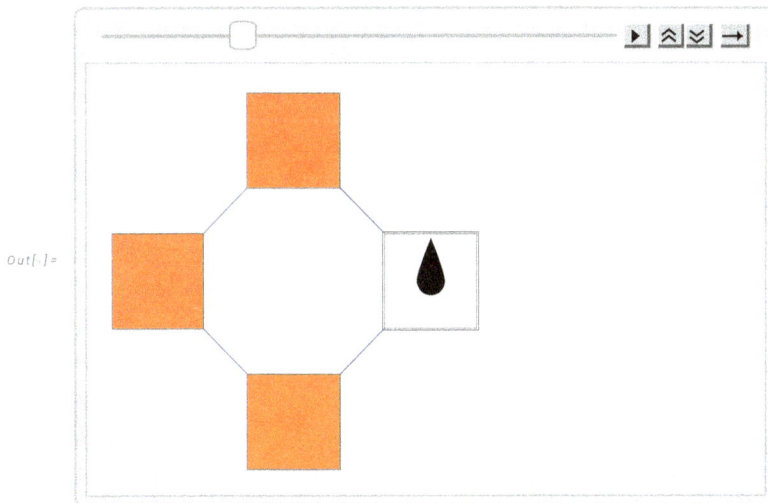

The rule 14 (1, 2)-TM travels around the cyclic graph, changing the color for each iteration. The behavior of the Turing machine can more conveniently be displayed by flattening out the graph to a 1D grid ordered by vertex number. This can be generalized for any graph type:

In[•]:= **EvolutionRulePlot[nrStates, nrColors, outEvolution]**

Out[•]=

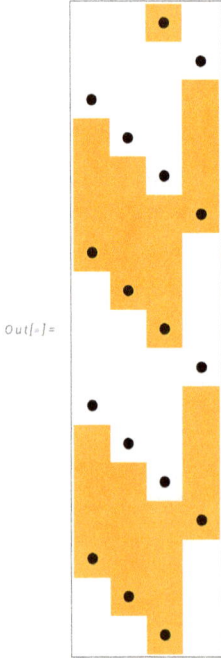

In the Turing machine plot shown previously, it is easy to see that the Turing machine repeats itself every eight steps. When running for many iterations, the Turing machine plot will be split into multiple panels to make the visualization easier. The cyclic behavior of rule 14 can also be illustrated using a state transition diagram:

In[•]:= **netTransitionDiagram[nrRule, inputGraph, nrStates, nrColors, nrEdges, nrIterations]**

Out[•]=

14

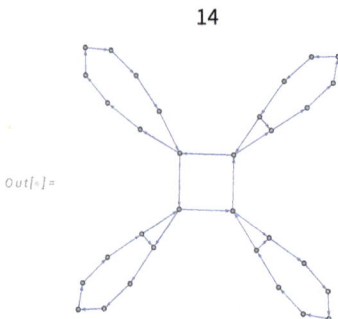

The state transition diagram shows how the Turing machine transitions from one state to the next, eventually looping around to the same state and repeating. For example, the first state transition in our current example is {1, {0, 0, 1, 0}, 3} -> {1, {0, 0, 0, 0}, 5}. The state diagram consists of the state transitions for all possible vertex starting positions, states and colors, which is why there are multiple loops in each diagram.

In this post we will also be analyzing the Turing machine's behavior based on the edge distance from the starting vertex points. For the cycle graph of four nodes, the max distance from the starting point is 2:

In[ ]:= **netGraphDistance[verticies, edges, outEvolution, startVertex, nrRule]**

Out[ ]:=

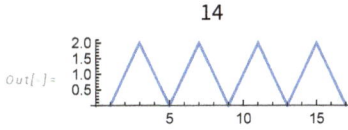

Cyclic Graph Turing Machine

This section shows the results for the cyclic graph Turing machine now expanded to seven vertices and multiple rules:

In[ ]:= **{adjacencyList, edges, verticies} = cyclicGraph[7, 1];**
nrEdges = 2;
inputGraph = adjacencyList;

In[ ]:= **SimpleGraph[Graph[verticies, edges], VertexLabels → Automatic]**

Out[ ]:=

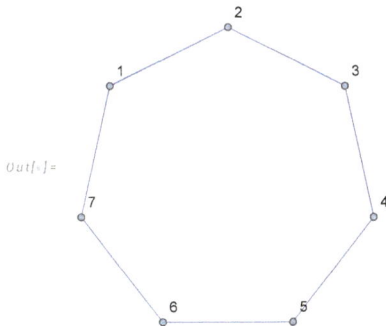

CyclicGraph[7]: 1-State and 2-Color Turing Machines

In[ ]:= **nrStates = 1;**
nrColors = 2;
nrIterations = 40;
startVertex = 5;
intialState =
{1, ReplacePart[Table[0, Length[inputGraph]], startVertex → 1], startVertex}

Out[ ]:= **{1, {0, 0, 0, 0, 1, 0, 0}, 5}**

Of the 16 possible rules for the (1, 2)-TM, rules 9, 10, 13 and 14 are worth highlighting. The Turing machine plot, state transition diagram and graph distance plot are shown here:

In[ ]:= **index = {9, 10, 13, 14};**

In[ ]:= **outEvolution = Map[NetTMEvolveList[NetTMRules[nrStates, nrColors, nrEdges, #],**
inputGraph , intialState , nrIterations] &, index];

In[ ]:= **Map[**
Labeled[EvolutionRulePlot[nrStates, nrColors, outEvolution[[#]], index[[#]], Top] &,
Range[Length[index]]] // Row[#, Spacer[10]] &
Map[netTransitionDiagram[#, inputGraph, nrStates, nrColors, nrEdges, nrIterations] &,
index] // Row[#, Spacer[10]] &
Map[netGraphDistance[verticies, edges, outEvolution[[#]], startVertex, index[[#]]] &,
Range[Length[index]]] // Row[#, Spacer[10]] &

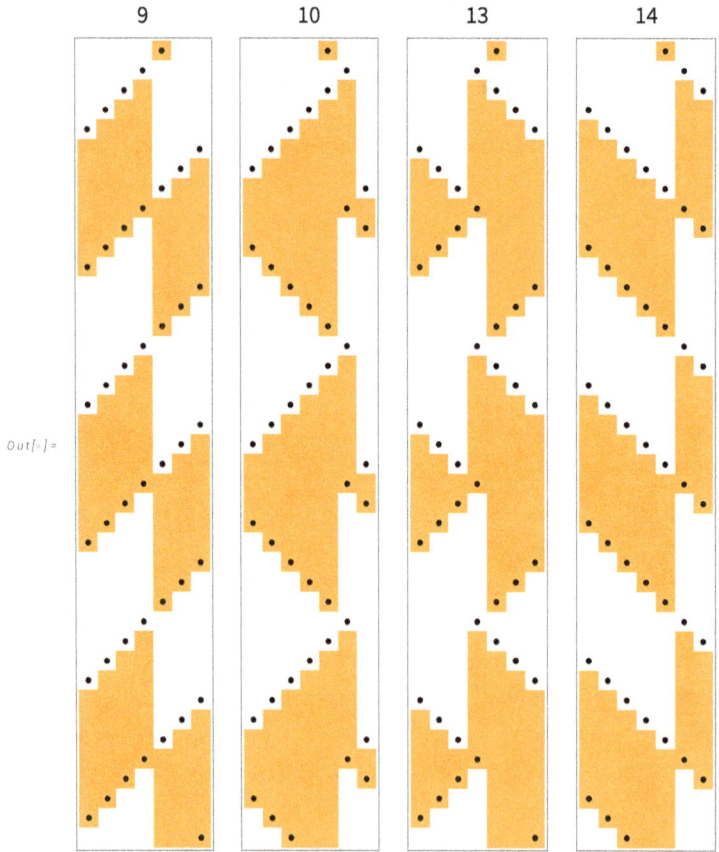

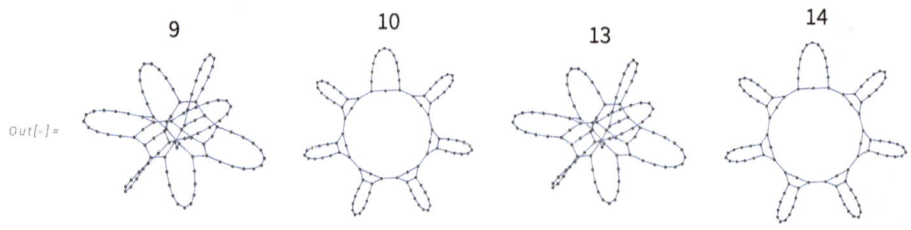

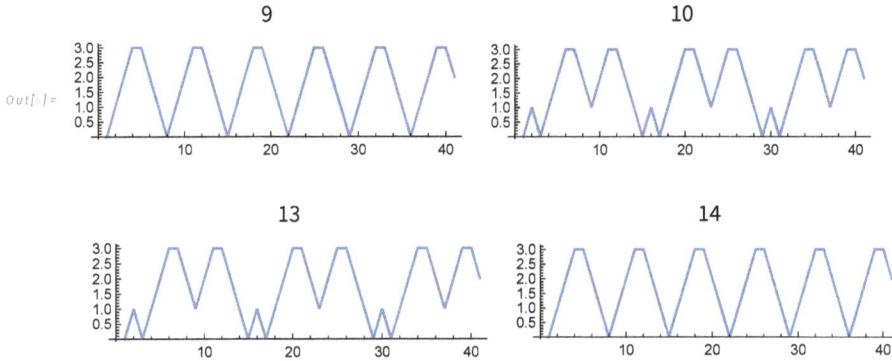

The four rules all show a repetitive pattern. Rules 9 and 13 (and 10 and 14) are almost mirror images of each other. However, there is a small difference in the traversal, which can be seen when comparing the edge distance diagram of 9 versus 13. Rules 10 and 13 both make a sidestep before continuing to cycle around the graph.

Besides rules 9, 10, 13 and 14, the other rules for the (1, 2)-TM on a cycle graph show only simple repetitive behavior, e.g. bouncing back and forth between two states. Despite their simplicity, the state transition diagrams can still appear quite complex for even the simplest scenarios.

CyclicGraph[7]: 2-State and 2-Color Turing Machines

To get more complex Turing machines, we increase the number of states to 2:

```
In[·]:= nrStates = 2;
        nrColors = 2;
        nrIterations = 300;
        startVertex = 5;
        intialState =
            {1, ReplacePart[Table[0, Length[inputGraph]], startVertex → 1], startVertex}
Out[·]= {1, {0, 0, 0, 0, 1, 0, 0}, 5}
```

For the 2-state and 2-color diagram cycle graph, here is the number of possible rules:

```
In[·]:= (2 × 2 × 2) ^ (2 × 2)
Out[·]= 4096
```

I have sifted through all of them and found that most of them have simple repetitive behavior, for example bouncing back and forth between two nodes. Rules 372, 316 and 3037 have been found to display more complex repetitive patterns:

```
In[·]:= index = {372, 316, 3037};

In[·]:= outEvolution = Map[NetTMEvolveList[NetTMRules[nrStates, nrColors, nrEdges, #],
            inputGraph, intialState, nrIterations] &, index];
```

In[•]:= `Map[Labeled[Table[Show[EvolutionRulePlot[nrStates, nrColors,`
`outEvolution[[#]][[i ;; i + 100 − 1]]], ImageSize → Medium],`
`{i, 1, nrIterations, 100}] // Row[#, Spacer[3]] &, index[[#]], Top] &,`
`Range[Length[index]]] // Row[#, Spacer[10]] &`

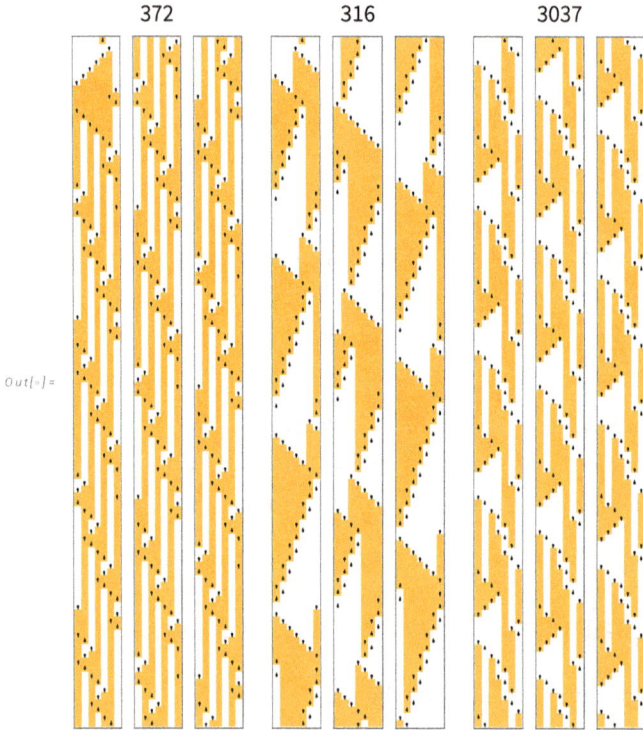

Out[•]=

In the previous figure, each panel is a continuation of the Turing machine for the respective rule. We see that there is a long repetitive pattern in each of the Turing machines. The repetitive pattern is also evident by looking at the graph distance of the Turing machine head, as shown here:

In[•]:= `Map[netTransitionDiagram[#, inputGraph, nrStates, nrColors, nrEdges, nrIterations] &,`
`index] // Row[#, Spacer[20]] &`
`Map[netGraphDistance[verticies, edges, outEvolution[[#]], startVertex, index[[#]]] &,`
`Range[Length[index]]] // Row[#, Spacer[20]] &`

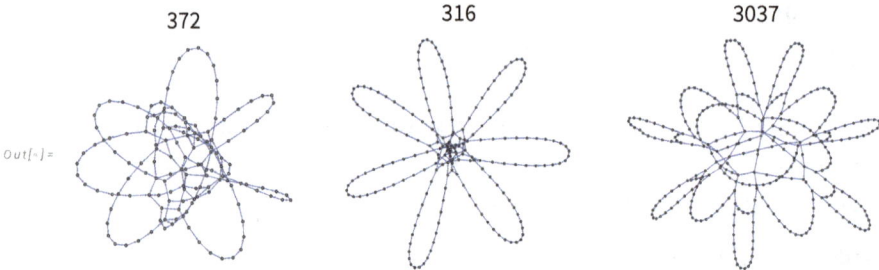

Out[•]=

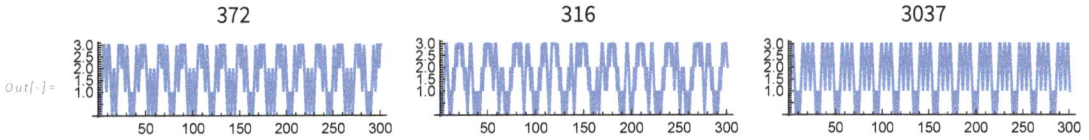

The pattern for rules 372, 316 and 3037 starts repeating itself after the first 54, 189 and 22 iterations, respectively:

```
In[·]:= Map[Length[DeleteDuplicates[outEvolution[[#]]]] &, Range[Length[index]]]
```

```
Out[·]= {54, 189, 22}
```

Some other interesting rules to explore for the (2, 2)-TM cycle graph are 1372, 3417, 1404, 2454, 1353, 3953, 2206 and 470.

Torus Graph Turing Machine

This section investigates the Turing machine on a torus. The torus is formed by labeling all the vertices on a 2D grid, then merging the edges from top to bottom and side to side. By merging the edges, we create a cyclic Turing machine that wraps around itself just like the cycle graph.

The torus formed by the 3×7 grid is created and shown in 1D. I have also run the simulations with other grid sizes of various permutations; however, I did not find that small changes had an impact on the Turing machine. When running on much larger grids, e.g. 30×70, the results are substantially different:

```
In[·]:= {adjacencyList, edges, verticies} = cyclicGraph[3, 7];
        nrEdges = 4;
        inputGraph = adjacencyList;
```

```
In[·]:= Graph[verticies, edges]
```

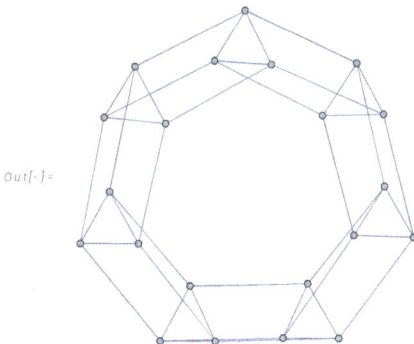

Torus[3, 7]: 1-State and 2-Color Turing Machines

```
In[•]:= nrStates = 1;
       nrColors = 2;
       nrIterations = 300;
       startVertex = 5;
       intialState =
           {1, ReplacePart[Table[0, Length[inputGraph]], startVertex → 1], startVertex}
```

Out[•]= {1, {0, 0, 0, 0, 1, 0, 0, 0, 0, 0, 0, 0, 0, 0, 0, 0, 0, 0, 0, 0, 0}, 5}

For the 1-state and 2-color torus, here are the number of possible rules:

```
In[•]:= (4 × 1 × 2) ^ (1 × 2)
```

Out[•]= 64

Like the 1-state, 2-color cyclic graph, most of the 64 rules for the torus are mirror images of each other, or otherwise simple repetitive behavior. Here is an example of a few of the more interesting rules:

```
In[•]:= index = {43, 44, 50};
```

```
In[•]:= outEvolution = Map[NetTMEvolveList[NetTMRules[nrStates, nrColors, nrEdges, #],
               inputGraph , intialState , nrIterations] &, index];
```

```
In[•]:= Map[Labeled[Table[Show[EvolutionRulePlot[nrStates, nrColors,
               outEvolution[[#]][[i ;; i + 100 – 1]]], ImageSize → Small],
           {i, 1, nrIterations, 100}] // Row[#, Spacer[3]] &, index[[#]], Top] &,
       Range[Length[index]]] // Row[#, Spacer[10]] &
```

Like with the (1, 2)-TM cycle graph, we see that rules 43 and 44 are near-mirror images of each other; however, they display different state transition diagrams:

```
In[ ]:= Map[netTransitionDiagram[#, inputGraph, nrStates, nrColors, nrEdges, nrIterations] &,
          index] // Row[#, Spacer[15]] &
       Map[netGraphDistance[verticies, edges, outEvolution[[#]], startVertex, index[[#]]] &,
          Range[Length[index]]] // Row[#, Spacer[15]] &
```

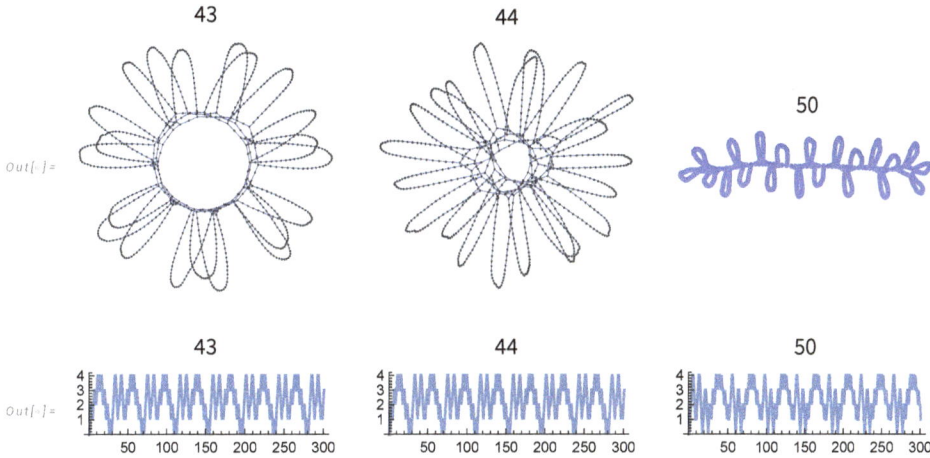

The pattern for rules 43, 44 and 50 starts repeating itself after the first 45, 45 and 49 iterations respectively:

```
In[ ]:= Map[Length[DeleteDuplicates[outEvolution[[#]]]] &, Range[Length[index]]]

Out[ ]= {45, 45, 49}
```

Torus (3 × 7): 2-State and 2-Color Turing Machines

The same torus formed by the 3 × 7 grid is used in this section. Now, we increase the complexity to two states. For the 2-state and 2-color torus, here is the number of possible rules:

```
In[ ]:= (4 × 2 × 2) ^ (2 × 2)

Out[ ]= 65 536
```

The number of rules has grown exponentially. During the project, I have run the Turing machine for about half the rules and found some that exhibit quite-complex behavior:

```
In[ ]:= nrStates = 2;
       nrColors = 2;
       nrIterations = 800;
       startVertex = 5;
       intialState =
          {1, ReplacePart[Table[0, Length[inputGraph]], startVertex → 1], startVertex};
```

In[ ]:= index = {23 281, 19 291};

In[ ]:= outEvolution = Map[NetTMEvolveList[NetTMRules[nrStates, nrColors, nrEdges, #],
      inputGraph , intialState , nrIterations] &, index];

In[ ]:= Map[Labeled[Table[Show[EvolutionRulePlot[nrStates, nrColors,
        outEvolution[[#]][[i ;; i + 100 − 1]]], ImageSize → Small],
      {i, 1, nrIterations, 100}] // Row[#, Spacer[3]] &, index[[#]], Top] &,
    Range[Length[index]]] // Row[#, Spacer[10]] &

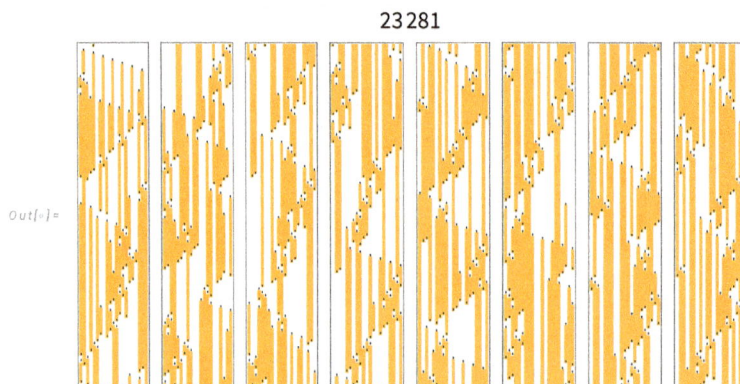

23 281

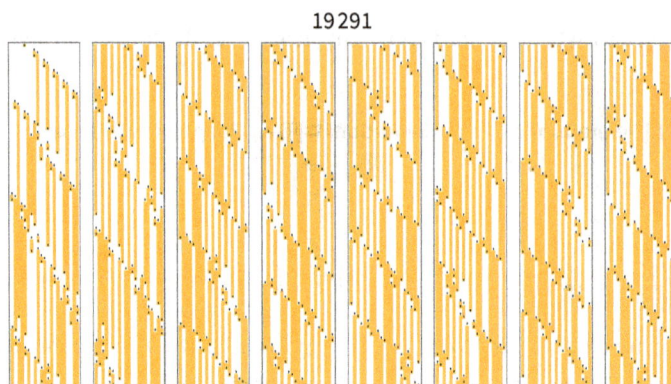

19 291

In[ ]:= Map[netGraphDistance[verticies, edges, outEvolution[[#]], startVertex, index[[#]]] &,
    Range[Length[index]]] // Row[#, Spacer[18]] &

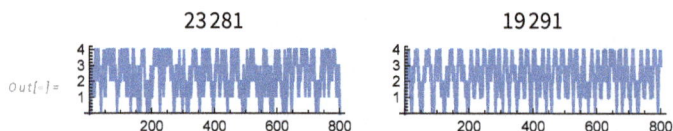

We see that the patterns are much more complex; however, they still display a certain repetitive cycling. Some other interesting rules to explore would be 3403, 2753, 3803, 63002, 19231, 19291, 23281, 31051, 31201 and 63811.

Large Torus (30 × 70): 2-State and 2-Color Turing Machines

With a larger system, the head of the Turing machine can travel much further from the starting node before wrapping around to the same position again. Using the same rules as in the previous section, we use a larger torus of 30 × 70 nodes:

```
In[ ]:= {adjacencyList, edges, verticies} = cyclicGraph[30, 70];
        nrEdges = 4;
        inputGraph = adjacencyList;
```

```
In[ ]:= Rasterize[Graph[verticies, edges]]
```

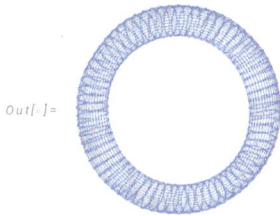

Out[ ]=

```
In[ ]:= nrStates = 2;
        nrColors = 2;
        nrIterations = 10 000;
        startVertex = 1000;
        intialState =
           {1, ReplacePart[Table[0, Length[inputGraph]], startVertex → 1], startVertex};
```

```
In[ ]:= index = {23 281, 19 291};
```

```
In[ ]:= outEvolution = Map[NetTMEvolveList[NetTMRules[nrStates, nrColors, nrEdges, #],
                         inputGraph , intialState , nrIterations] &, index];
```

```
In[ ]:= Map[netGraphDistance[verticies, edges, outEvolution[[#]], startVertex, index[[#]]] &,
          Range[Length[index]]] // Row[#, Spacer[20]] &
```

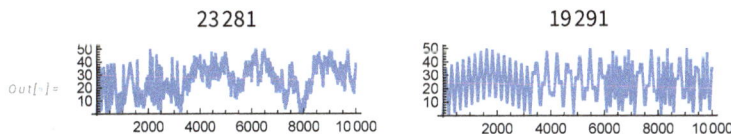

Out[ ]=

The graph distance shown for rule 23281 looks like stochastic behavior and could have been produced by stock price data. The perceived randomness from rule 23281 is amazing considering that it is the result of a fully deterministic process that repeats cyclically. Rule 19291, on the other hand, cycles regularly through the entire torus in the beginning and then begins a more chaotic cycle.

The Turing machine plots for these two rules are shown here. Both graphs start out with all the nodes colored white, and as the Turing machine travels along the edges, the nodes are gradually colored orange. For a further investigation, it would be interesting to highlight the edges on the torus that the Turing machine travels around:

In[◦]:= Map[Labeled[Rasterize[Show[EvolutionRulePlot[nrStates, nrColors, outEvolution[[#]]],
ImageSize → Medium], index[[#]], Top] &,
Range[Length[index]]] // Row[#, Spacer[20]] &

23 281　　　　　19 291

Out[◦]=

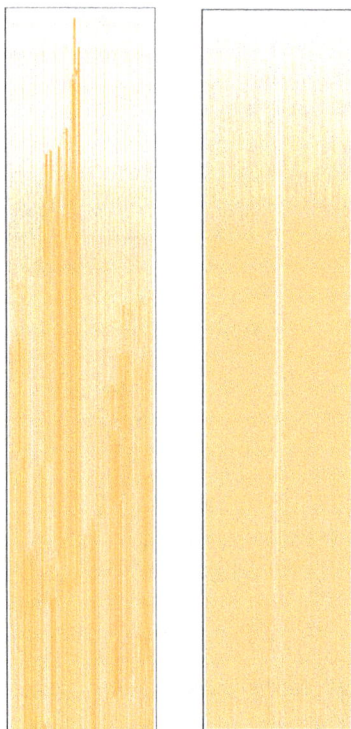

Turing Machine on the Sierpiński Graph

In this section, we generate the Turing machine on the Sierpiński network. The traditional
Sierpiński graph has the property that the corners of the vertices only have three edges, while
the rest of the nodes have four. To implement the Turing machine on the Sierpiński graph,
we need to deal with the corner effects. In this project, we have generated a Sierpiński graph
by taking another Sierpiński graph of one degree smaller, merging the three corners. The two
Sierpiński graphs that are joined are shown here:

In[◦]:= Map[Rasterize[MeshConnectivityGraph[SierpinskiMesh[#]]] &, {1, 2}] //
Row[#, Spacer[15]] &

Out[◦]=

By merging the three corners of the two Sierpiński graphs, we get a cyclic Sierpiński graph
that wraps around and has all vertices with four edges.

With the torus and the cycle graph, there is a natural left, right, up and down that can be defined. The geometry of the Sierpiński makes the decision of which node is up and which down more difficult. During the project, we explored various options, from simply numbering the vertices to treating each neighboring triangle as a ternary diagram. In the end, we decided to assign a coloring to each of the edges. The decision of left vs. right thus becomes red, yellow, purple, orange.

The geometry of the Sierpiński graph has been generalized to any degree; however, in the rest of this section, we will focus on the smallest Sierpiński of degree one, shown here:

```
In[ ]:= g = SierpinskiNetwork[1, VertexLabels → Automatic, ImageSize → Medium];
        c = ResourceFunction["EdgeColoring"][g, Method → "Optimum"];
        Rasterize[
          HighlightGraph[Graph[g, EdgeStyle → Thickness[0.02], VertexSize → 0.2], c]]
        adjacencyList = Table[Table[AdjacencyList[c[[colors]], vertex], {colors, 1, 4}] // Flatten,
              {vertex, 1, VertexCount[g]}];
        verticies = Range[VertexCount[g]];
        edges = Flatten[c];
        nrEdges = 4;
        inputGraph = adjacencyList;
```

Out[ ]=

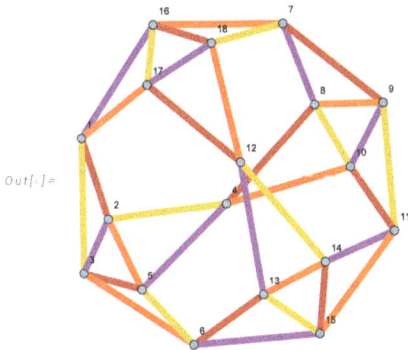

Sierpiński Network with 1-State and 2-Color Turing Machines

```
In[ ]:= nrStates = 1;
        nrColors = 2;
        nrIterations = 300;
        startVertex = 5;
        intialState =
            {1, ReplacePart[Table[0, Length[inputGraph]], startVertex → 1], startVertex}
```

Out[ ]= {1, {0, 0, 0, 0, 1, 0, 0, 0, 0, 0, 0, 0, 0, 0, 0, 0, 0, 0}, 5}

```
In[ ]:= index = {59, 41};
```

In[•]:= `outEvolution = Map[NetTMEvolveList[NetTMRules[nrStates, nrColors, nrEdges, #],`
`inputGraph , intialState , nrIterations] &, index];`

In[•]:= `Map[Labeled[Table[Show[EvolutionRulePlot[nrStates, nrColors,`
`outEvolution[[#]][[i ;; i+100-1]]], ImageSize → Medium],`
`{i, 1, nrIterations, 100}] // Row[#, Spacer[3]] &, index[[#]], Top] &,`
`Range[Length[index]]] // Row[#, Spacer[10]] &`

Out[•]=

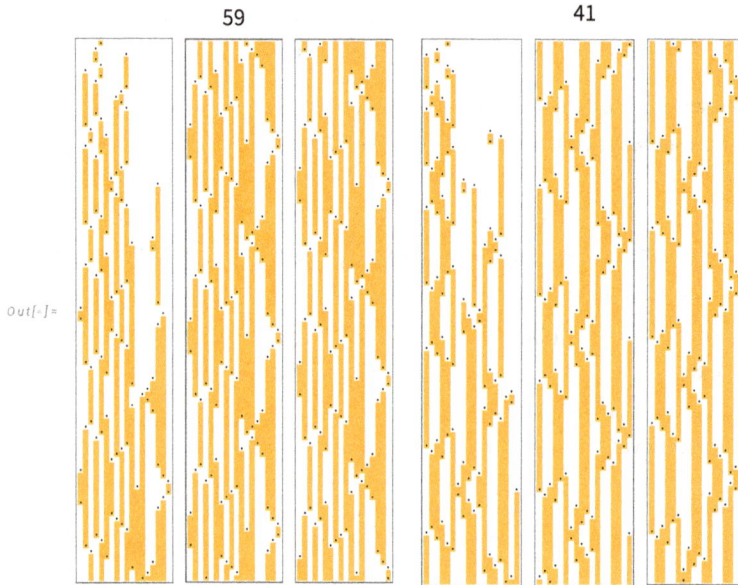

In[•]:= `Map[netTransitionDiagram[#, inputGraph, nrStates, nrColors, nrEdges,`
`nrIterations] &, index]`

Out[•]=

In[•]:= `Map[netGraphDistance[verticies, edges, outEvolution[[#]], startVertex, index[[#]]] &,`
`Range[Length[index]]] // Row[#, Spacer[15]] &`

Out[•]=

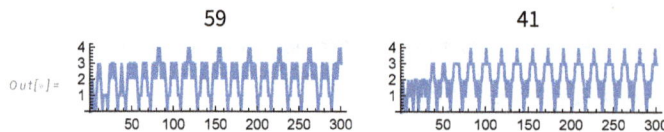

We immediately see that the Sierpiński (1, 2)-TM has more complex and varied behavior compared to what we saw for the cycle graph and the torus. The state transition diagram especially shows a different structure, which wasn't observed in any of the underlying rules for the torus.

Sierpiński Network with 2-State and 2-Color Turing Machines

As with the cycle graph and the torus, we now investigate the 2-state and 2-color Turing machine for the Sierpiński network:

```
In[ ]:= nrStates = 2;
       nrColors = 2;
       nrIterations = 1000;
       startVertex = 5;
       intialState =
          {1, ReplacePart[Table[0, Length[inputGraph]], startVertex → 1], startVertex};
```

```
In[ ]:= index = {50 001, 59 656};
```

```
In[ ]:= outEvolution = Map[NetTMEvolveList[NetTMRules[nrStates, nrColors, nrEdges, #],
               inputGraph , intialState , nrIterations] &, index];
```

```
In[ ]:= Map[Labeled[Table[Show[EvolutionRulePlot[nrStates, nrColors,
                  outEvolution[[#]][[i ;; i + 100 − 1]]], ImageSize → Small],
              {i, 1, nrIterations, 100}] // Row[#, Spacer[3]] &, index[[#]], Top] &,
          Range[Length[index]]] // Row[#, Spacer[10]] &
       Map[netGraphDistance[verticies, edges, outEvolution[[#]], startVertex, index[[#]]] &,
          Range[Length[index]]] // Row[#, Spacer[20]] &
```

50 001

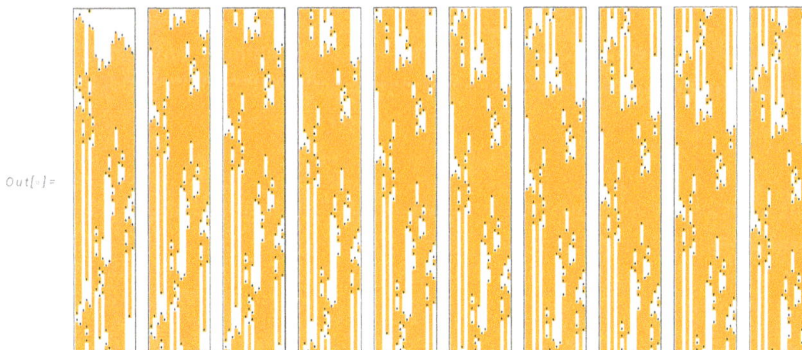

59 656

50 001   59 656

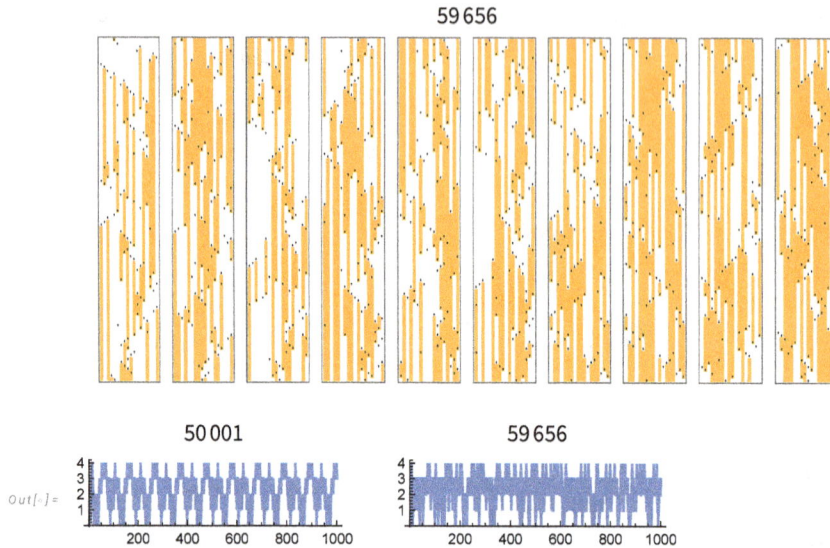

With the previously shown Turing machines, the use of the state transition diagram becomes much too big to display. Looking at rule 50001, we see it has a repetitive structure in the edge distance plot. The coloring in the Turing machine plot for rule 50001 is quite different from most of the other Turing machines explored, being mostly colored orange with some holes of white here and there. With rule 59656, there is not any obvious repetitive behavior. I found from running it for 10 thousand iterations that there is a repetitive cycle that emerges after 5326 iterations.

The complexity of these Turing machines starts to exceed what can be discerned from visual inspection. One test we can perform to understand the behavior is to see if the Turing machine visits all nodes (completes a Hamiltonian cycle).

From the following code, we see that both the Turing machines do in fact visit all 18 nodes within the 10 thousand iterations:

```
In[·]:= Map[Length[Tally[Table[outEvolution[[#]] [[c]][[3]] , {c, 1, Length[outEvolution[[#]]]}] ]] &,
        Range[Length[index]]]
```

```
Out[·]= {18, 18}
```

Some other rules worth exploring are rule numbers 18785, 51736, 60181, 40051, 59446, 48706, 40516, 60166 and 59656.

Extended Sierpiński Network with 2-State and 2-Color Turing Machines

To get the image used for the title page, the (2, 2)-TM with rule 59656 is simulated for five thousand iterations and shown:

```
In[*]:= nrStates = 2;
        nrColors = 2;
        nrIterations = 5000;
        startVertex = 5;
        intialState =
          {1, ReplacePart[Table[0, Length[inputGraph]], startVertex → 1], startVertex};
```

```
In[*]:= index = {50 001, 59 656};
```

```
In[*]:= outEvolution = Map[NetTMEvolveList[NetTMRules[nrStates, nrColors, nrEdges, #],
                inputGraph , intialState , nrIterations] &, index];
```

```
In[*]:= Labeled[EvolutionRulePlot[nrStates, nrColors, outEvolution[[2]]], index[[2]], Top]
```

59 656

Bonus: Machine Music

```
In[ ]:= g = SierpinskiNetwork[1, VertexLabels → Automatic, ImageSize → Medium];
       c = ResourceFunction["EdgeColoring"][g, Method → "Optimum"];
       adjacencyList =
           Table[Table[AdjacencyList[c[[colors]], vertex], {colors, 1, 4}] // Flatten,
               {vertex, 1, VertexCount[g]}];
       verticies = Range[VertexCount[g]];
       edges = Flatten[c];
       nrEdges = 4;
       inputGraph = adjacencyList;
```

One area explored during the project was the repetitive structure of the Turing machines by converting the machines into sound. In this section, I use the Sierpiński network for the (1, 2)-TM:

```
In[ ]:= nrStates = 1;
       nrColors = 2;
       nrIterations = 100;
       startVertex = 5;
       intialState =
           {1, ReplacePart[Table[0, Length[inputGraph]], startVertex → 1], startVertex};
```

```
In[ ]:= index = {59, 41};
```

```
In[ ]:= outEvolution = Map[NetTMEvolveList[NetTMRules[nrStates, nrColors, nrEdges, #],
               inputGraph , intialState , nrIterations] &, index];
```

To generate a pleasant melody, I assign each node in the network to a note in the C major scale. We start the scale one octave below middle C, then we iteratively add more keys over multiple octaves:

```
In[ ]:= scale = Flatten[
               NestList[12 + # &, {-12, -10, -8, -5, -3}, Ceiling[Length[inputGraph] / 5] - 1]];
       scale = Take[scale, Length[inputGraph]]
```

```
Out[ ]= {-12, -10, -8, -5, -3, 0, 2, 4, 7, 9, 12, 14, 16, 19, 21, 24, 26, 28}
```

Here is the melody for rules 59 and 41. The corresponding Turing machine plot is shown here:

```
In[ ]:= Map[Labeled[Sound[SoundNote[#] &/@
            (Pick[scale, #, 1] &/@Table[outEvolution[#][n][2], {n, 1, nrIterations}])],
          index[#], Top] &, Range[Length[index]]]
```

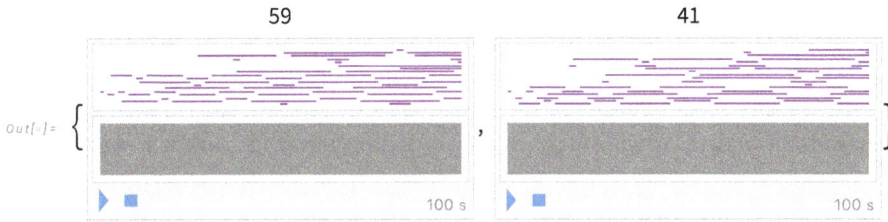

```
In[ ]:= Map[Labeled[Rasterize[Show[EvolutionRulePlot[nrStates, nrColors, outEvolution[#]],
            ImageSize → Medium], index[#], Top] &,
          Range[Length[index]]] // Row[#, Spacer[20]] &
```

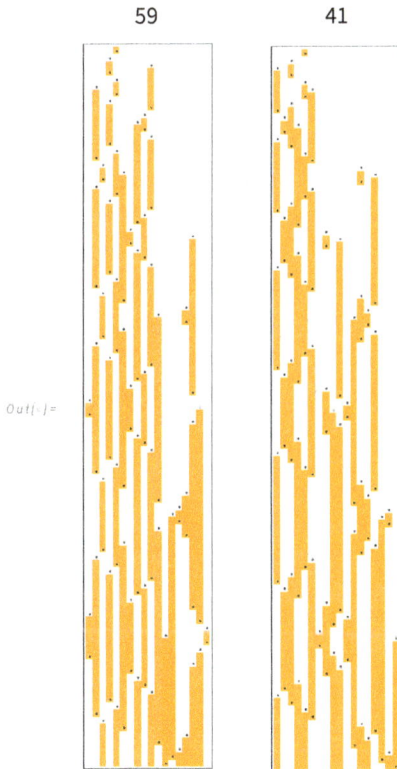

The rhythm used for these examples is very monotonous, since each key is pressed down on every beat. The algorithm could be improved by having the keys only being pressed down a single time when the node changes from white to orange. This would generate a much more dynamic and musical composition.

Future Work

In this project, I have only scratched the surface for the exploration of Turing machines on graphs.

- The cyclic behavior of the Turing machines studied needs to be investigated further to understand their underlying functional behavior, for example whether a rule is a binary counter or if it cycles through a set of primary numbers.

- It needs to be determined if the coloring scheme used for the Sierpiński graph is isomorphic and what the impact would be on the results for the Turing machine if it isn't.

- There are endless possibilities of various graphs that could be explored. For future work, I would propose that the hexagonal torus and Petersen graphs be explored as examples of graphs with a valence of 3.

Conclusion

This post has investigated the 1-state, 2-state and 2-color Turing machines on finite graphs. The graphs used were the 1D cyclic graph, the torus and the cyclic Sierpiński network. It was found that the cyclic Sierpiński network showed more complex patterns for the (1, 2)-TM than the cycle graph and torus. Similarly, the (2, 2)-TM Sierpiński network showed more complexity as compared to the (2, 2)-TM on the torus and cycle graph.

However, when running the Turing machine on a larger torus (30×70) with more room for the head to move around, I found that there was a stochastic-looking behavior that arises. The perceived stochastic behavior is impressive considering that it is fully deterministic and follows a regular cyclic pattern.

The Turing machines on the Sierpiński network are worth further exploration. Specifically, it is necessary to prove that the coloring scheme used for the Sierpiński network is isomorphic. A deeper analysis should also be performed on the Sierpiński network of higher degree and on the torus with more nodes.

Acknowledgments

- Thank you to my mentors Robert Nachbar and Alejandra Ortiz for guiding me through the project. Specifically, Alejandra came up with the idea for coloring the Sierpiński graph to determine the coordinate system and Robert came up with the generalized geometry for the cyclic Sierpiński graph.

- Thank you to Stephen Wolfram for the initial project idea and discussions on how best to analyze the results.

- Thank you to all the Wolfram staff who helped organize the 2023 Wolfram Summer School.

References

1. S. Wolfram (2002), *A New Kind of Science*, Wolfram Media, Inc.

2. J. Stocke (2023), "Turing Machine Evolution on Graphs Using Principles of Multiway Computation," *Wolfram Community*. community.wolfram.com/groups/-/m/t/2946777.

Access the Full Code

Scan or visit wolfr.am/WSS2023-Graham.

Cite This Notebook

"Turing Machines on Graphs"

by Stefan Graham

Wolfram Community, STAFF PICKS, July 12, 2023

community.wolfram.com/groups/-/m/t/2959123

Study the Effect of Small Changes in Large Language Models

LUHAN CHENG

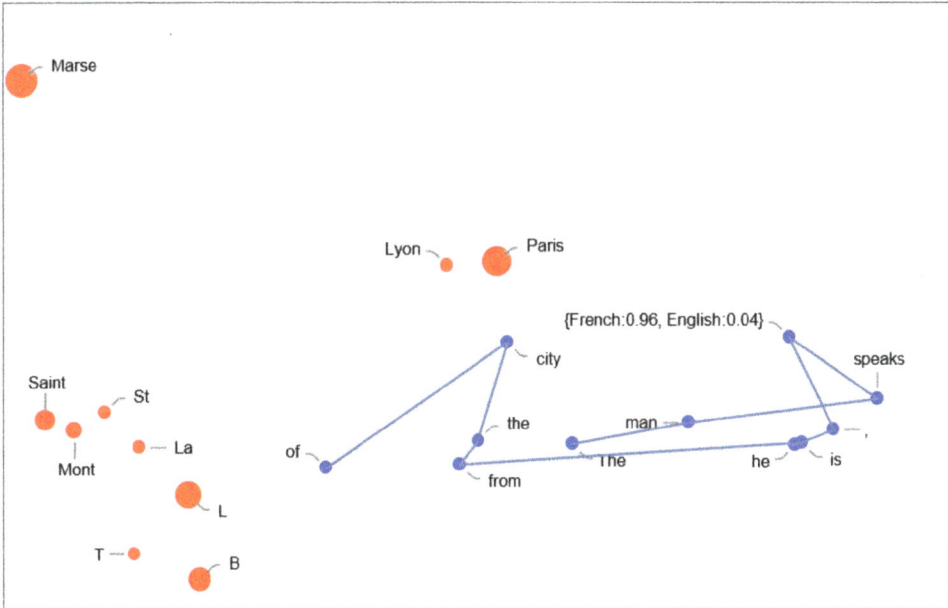

Large language models (LLMs) have demonstrated their strong capabilities in language modeling, text generation and many other natural language understanding tasks compared to their predecessors. We hypothesize LLMs gain their strength by successfully mapping words onto the semantic space and traversing through the semantic space with a high degree of stability. We model the computation of LLMs as trajectories in their embedding spaces. We study the stability of the trajectory by introducing small amounts of controlled noise into the input word embeddings. We demonstrate that GPT-3 is more capable of resisting small perturbations in the input space compared to the GPT-2 model.

We start with investigating the GPT-2 model [1] provided by the Wolfram Neural Net Repository. The model is pre-trained on WebText data. We will use a version with 117 million parameters for our experiments, though the same workflow can easily be extended to versions with a larger parameter size.

The input sentence is first converted into a sequence token. The stream of this token is then converted to an $n \times 768$ matrix through both a positional embedding layer and a token embedding layer. The output from the embedding layer is then fed into the decoder module, which is the major component of the model. It consists of 12 self-attention modules that independently read/write information from/to the residual stream. The output of the decoder is an $n \times 768$ matrix. We feed the last row of this matrix to the linear classifier layer, which outputs the probability distribution over tokens through a softmax layer:

In[·]:= **gpt2lm**

Out[·]= NetChain[▣ ▮▮ Input port: string / Output port: class]

Similarity Matrix for Concept Space Mapping

To gain an understanding of the structure of the semantic space, we first explore what the similar embeddings are in this space. Note that it is impossible to enumerate all possible token sequences; therefore, we choose to take a specific example that will be used later in this notebook.

The successive phrases for a sentence, "The man speaks French, he is from the city of," is defined as the following:

In[·]:= **Column[StringJoin /@ foldTokens @ "The man speaks French, he is from the city of"]**

The
The man
The man speaks
The man speaks French
The man speaks French,
Out[·]= The man speaks French, he
The man speaks French, he is
The man speaks French, he is from
The man speaks French, he is from the
The man speaks French, he is from the city
The man speaks French, he is from the city of

We then feed each of these successive phrases into GPT-2 and GPT-3 (text-embedding-ada-002) to obtain the 768-dimensional embedding vector for GPT-2 and the 1,536-dimensional vector for GPT-3. We then calculate the pairwise cosine similarity between them and plot them as a heat map:

In[·]:= **With[{sentence = "The man speaks French, he is from the city of"},**
 Row[{subsentenceSimilarity[sentence, PlotLabel → "GPT2", ImageSize → 330],
  subsentenceSimilarityGPT35[sentence, PlotLabel → "GPT3", ImageSize → 330]}]]

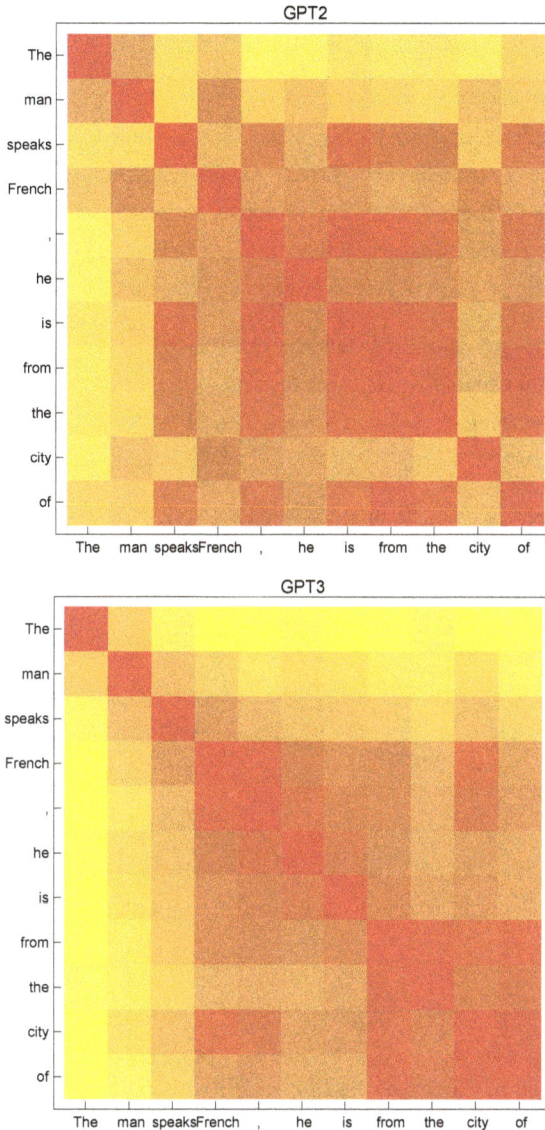

It is interesting to see the segregated structure shown in GPT-3's similarity plot. After the model sees the token "French", every subsequent phrase is comparatively similar. The same situation happens when the model sees the token "from". The similarity matrix plot for GPT-2 did not demonstrate the same pattern. A few more examples with different sentences are shown at the end of the notebook.

We hypothesize that the GPT-3 model implicitly learns to segregate the embedding space into hierarchies of contextualized spaces, where it learns to progressively switch to more specific contexts following the input sequence. A future, more detailed study is needed to confirm our hypothesis.

Now, what are the small changes we can make in the embedding layer? Our solution is to add scaled unit normal noise.

Random Noise at Embedding

We first experiment with adding random noise to the embedding and observe the output distribution. The random noise is generated from the unit normal distribution and scaled according to the noise level α. The following plot shows the cumulative distribution of the output with respect to different α values. Tokens on the x axis are sorted in descending order according to their probabilities. Each plot is averaged over 10 trials.

It seems the model is more confident when there is more noise introduced into the embedding. The confidence level does not scale linearly with respect to the level of noise added to the embedding. There is a gap between $\alpha = 0.8$ and $\alpha = 1.4$. The GPT-2 model seems to be resistant to noise level up to $\alpha = 0.4$:

`In[ ]:=` **plotSentencePerturb[Range[0, 2, 0.2], 10, ImageSize → Large,**
        PlotLabel → "Cumulative Distribution Plot of the Output"]

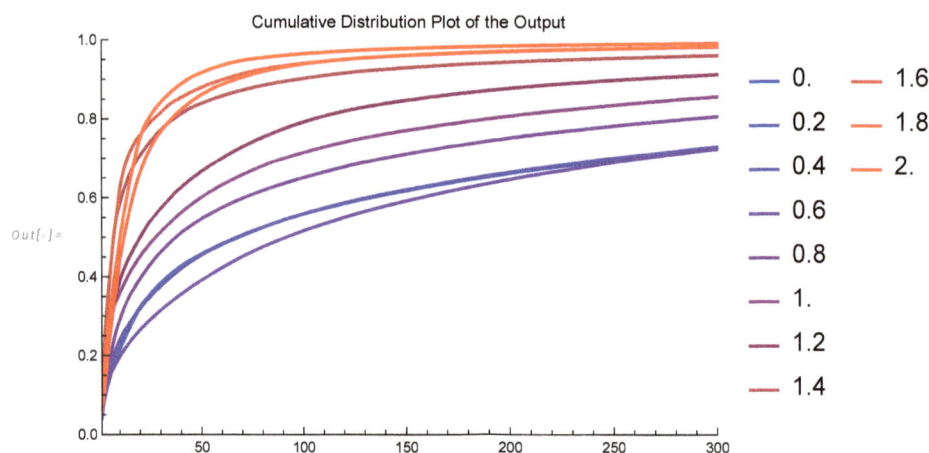

It is interesting to study the embedding space of the sentence; however, it also comes with a few issues. Firstly, the noise added does not go through the decoder, which is the most important component of the model. Also as important is that there isn't a good way to propagate the noise signal back to the neural network because it can only take a sequence of tokens as inputs. Therefore, we decided it would be more interesting to take a look at the perturbation at the word-embedding level.

Footnotes:

- Our method of introducing unit normal noise into the embedding layer is out of convenience. A more interesting approach would be to first obtain a PCA reduction and then apply scaled noise to each principle component.

- Each tick in the x axis does not correspond to the same token because the top three hundred predictions are most likely varying. It would be interesting to investigate further the question of at which point the network starts to assign high probability to a meaningless next token.

Interpolating Token Embeddings

There are a couple of advantages to interpolating token embeddings. The spaces of the inputs are drastically reduced to 50 thousand tokens, and more importantly, distance measurements are an intrinsic part of human perception. We can immediately tell that the words *dog* and *cat* are semantically closer than *dog* and *spacecraft*. Therefore, we can control the change introduced into the token-embedding space by interpolating between token-embedding vectors.

We have also measured the Kullback–Leibler (KL) divergence between the output distributions throughout the interpolation. We fix an origin token, in this case "French", and interpolate it to different tokens. Some of them are semantically closer in human perception, while others are further away:

```
interpolationDivergence[" French",
    {" tired", " she", " tall", " Italian", " English", " Spanish"}, Range[0, 1, 0.1],
    StringTemplate["The man speaks`` , he is from the city of"],
    PlotStyle → {Red, Blue, Black, Pink, Yellow, Green}]
```

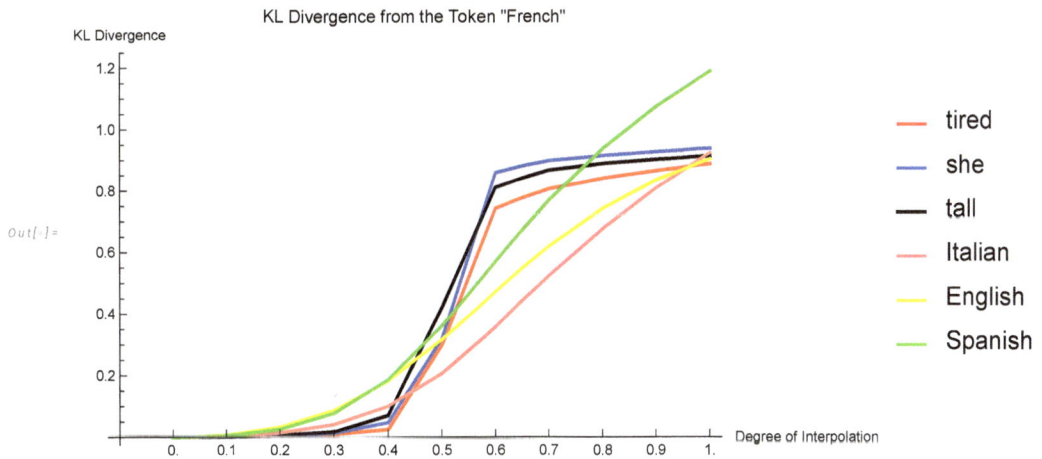

There seems to be a general trend that semantically closer (under human perception) tokens result in smoother output distributions by the model. One possible interpretation is that models are taught to always stick with the semantic, meaningful regions in the embedding space. If the interpolation leads to traveling between two semantically meaningful spaces, then it would be best to quickly travel through the semantically meaningless region in the middle, which leads to the seemingly sudden transition in the output distribution.

Now let's take a more detailed look at the transition between semantic spaces. For a pair of tokens, in this example "French" and "English", we can calculate the interpolated token-embedding vector and then feed it into the model to obtain the sentence-level embedding vector. We then calculate the pairwise cosine similarity for all sentence-level embedding vectors.

The left-hand figure represents the interpolation between the tokens "French" and "tired", and the right-hand figure represents the interpolation between the tokens "French" and "English". It is interesting that we can see a sharper transition between the conceptual spaces of "French" to "tired" than from "French" to "English". It is an indicator that there may be discontinuation from the concept space of "French" to "tired":

```
In[ ]:= Row[{
        plotSimilarity[Range[0, 1, 0.1], interpToken[Range[0, 1, 0.1], {" French", " tired"}],
          PlotLabel → "Interpolation between token French and tired", ImageSize → 330],
        plotSimilarity[Range[0, 1, 0.1], interpToken[Range[0, 1, 0.1], {" French", " English"}],
          PlotLabel → "Interpolation between token French and English", ImageSize → 330]
      }]
```

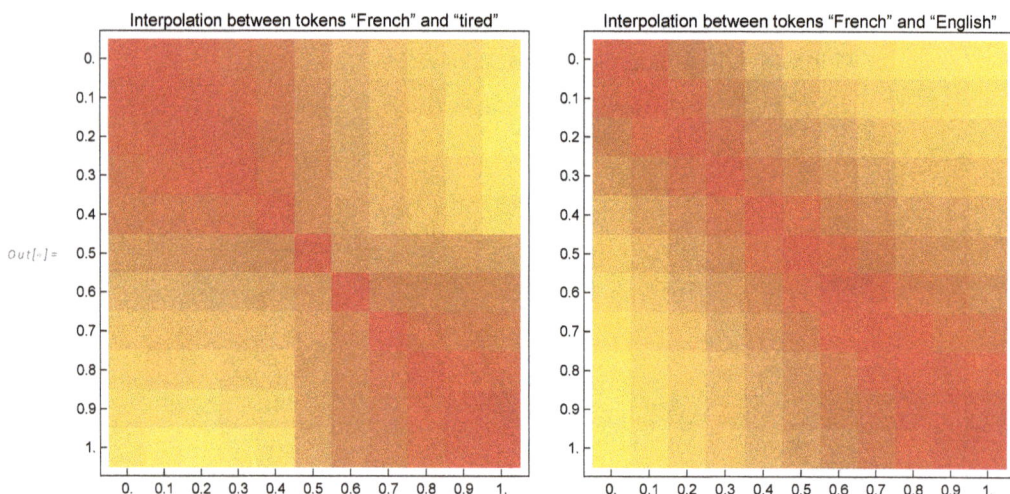

Footnotes:

- We used linear interpolation out of convenience, but it is likely there are privileged bases in the embedding space, so a nonlinear interpolation that involves first identifying these privileged bases might be more appropriate.

- The choice of tokens/sentences used in the first figure is mostly for the convenience of the rest of the notebook. But there is no reason why you cannot try different kinds of token/sentence combinations. In fact, a more rigorous study would involve sampling from all types of words using the WordList function and calculating their averaged KL divergences through interpolation.

- Additional examples are provided at the end of the notebook.

Trajectories in the Embedding Space

Now let's try to get a global view on how the GPT model moves in the embedding space.

Here, we show the trajectories of GPT-2 in its embedding space. Each blue point corresponds to an embedding vector of one of the successive phrases. The red spheres correspond to the 10 most likely tokens predicted by GPT-2, with their sizes scaled according to the probabilities; their embeddings are also whole-sentence embeddings to make them comparable with the trajectory embeddings. The trajectory corresponds to the sentence "The man speaks < slot >, he is from the city of." Each of the four figures corresponds to a different language inserted at < slot >.

In general, GPT-2 seems to be making quite sensible decisions. For example, when the input word is "French", it correctly predicted "Marse" and "Paris". And in most cases, there are clear separations between character tokens and whole-word tokens.

The general trajectory of GPT-2 does seem to be relatively stable under linear transformations. However, the relative position between the predicted embedding and the trajectory's embedding causes the trajectory plot to look chaotic:

```
In[ ]:=  Grid@Map[plotTrajectory3D[#, ImageSize → 330] &,
           Map[StringTemplate["The man speaks ``, he is from the city of"],
            {{"English", "French"}, {"German", "Spanish"}}, {2}], {2}]
```

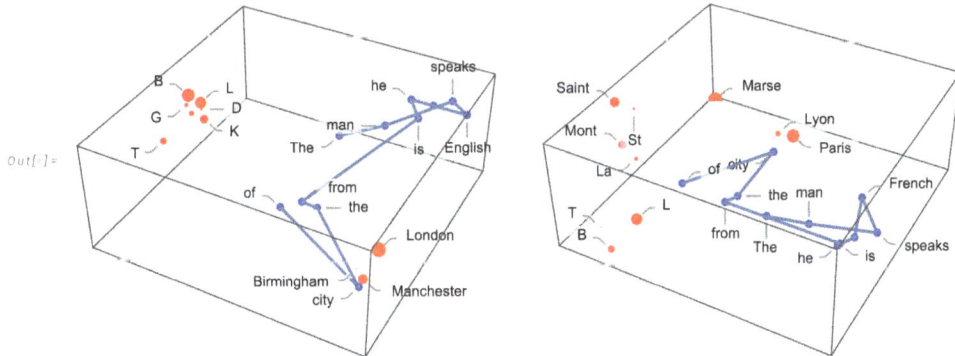

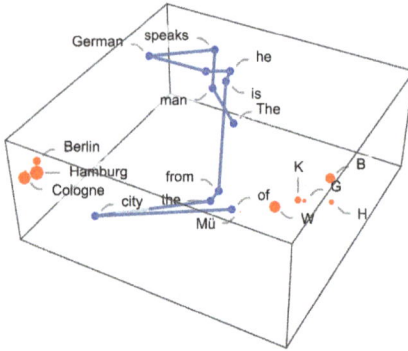
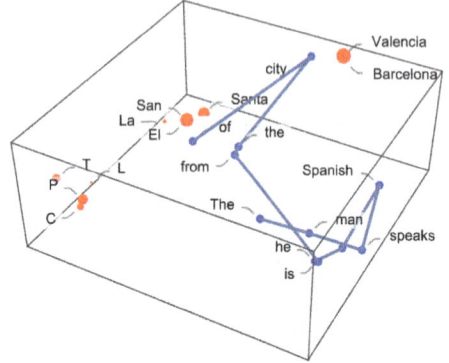

However, for GPT-3, the trajectory is much more stable. The perturbation did not change the general direction of the trajectory.

And compared to GPT-2, GPT-3 seems to provide better contextualization. For example, the embedding of the second "he" is quite far away from the first "he" for GPT-3, but the same cannot be said for GPT-2. This also matched our observation in the similarity heatmap example:

```
In[·]:= Grid@Map[plotTrajectoryGPT353D[♯, ImageSize → 330] &,
          Map[StringTemplate["The man speaks ``, he is from the city of"],
          {{"English", "French"}, {"German", "Spanish"}}, {2}], {2}]
```

Out[·]=

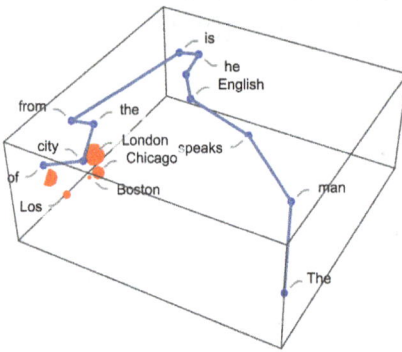
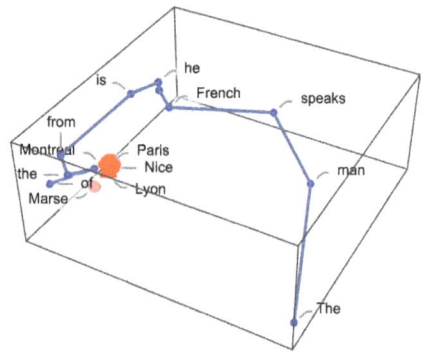

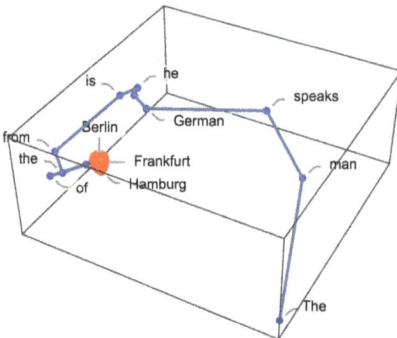
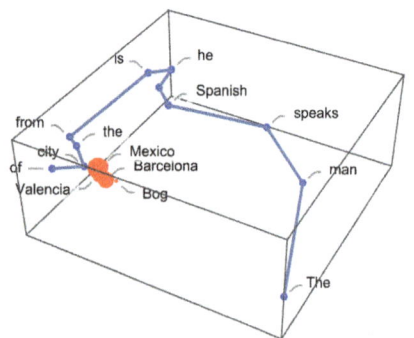

The comparison is even more obvious when we use a more complicated sentence:

```
In[·]:= Grid@Map[plotTrajectory3D[♯, ImageSize → 330] &,
         Map[StringTemplate["He is from the city of ``, he speaks the language of the"],
         {{"London", "Paris"}, {"Berlin", "Madrid"}}, {2}], {2}]
```

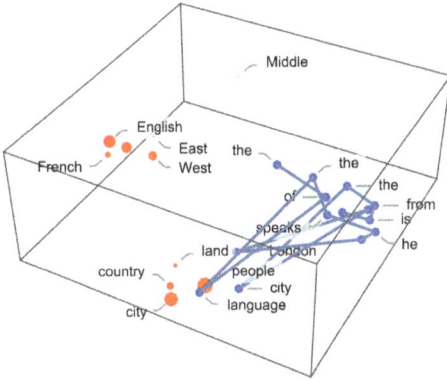 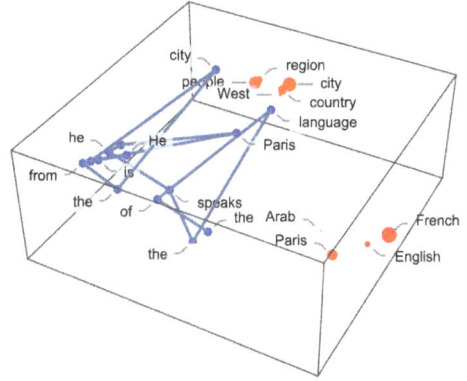

Out[·]=

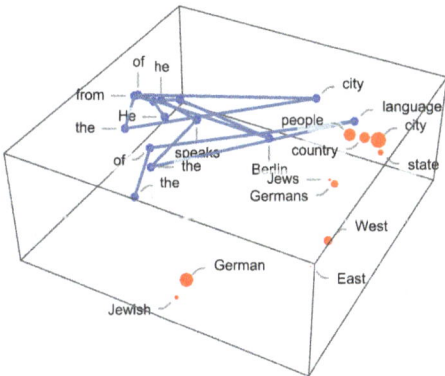 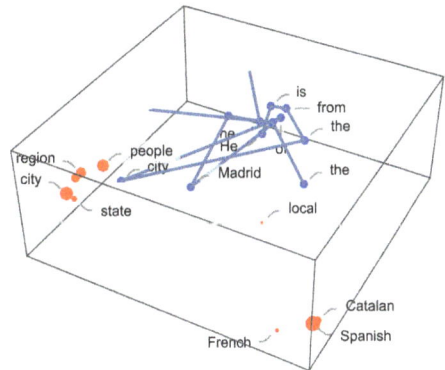

```
In[·]:= Grid@Map[plotTrajectoryGPT353D[♯, ImageSize → 330] &,
         Map[StringTemplate["He is from the city of ``, he speaks the language of"],
         {{"London", "Paris"}, {"Berlin", "Madrid"}}, {2}], {2}]
```

Out[·]=

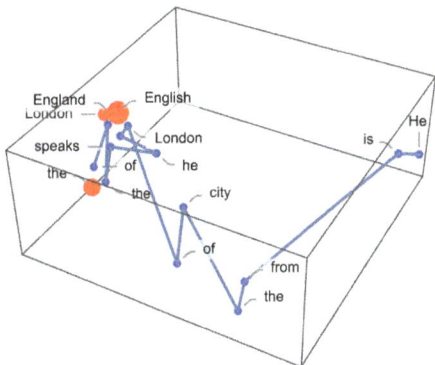 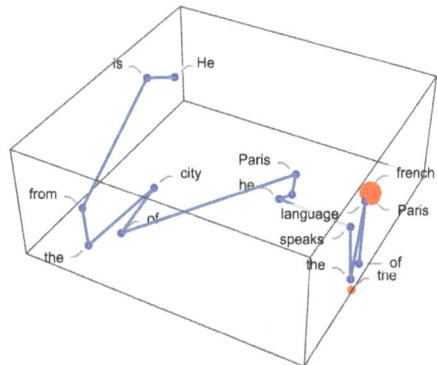

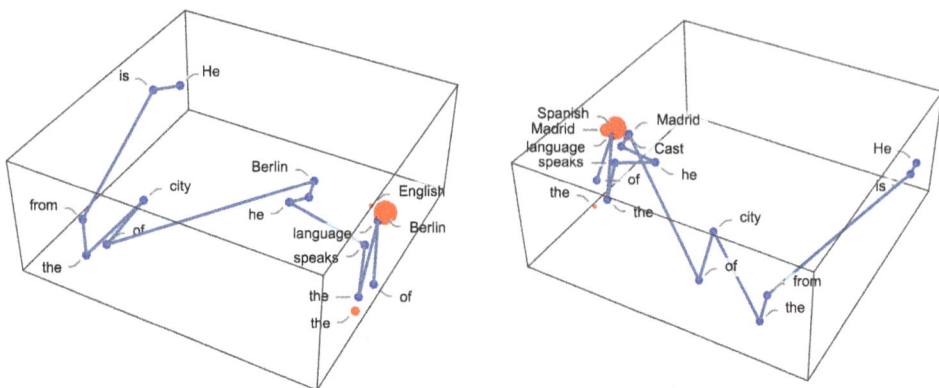

The general trajectory of GPT-2 seems to break down completely when we use meaningless characters to fill the slot, but the trajectory of GPT-3 seems to remain stable:

```
In[ ]:=  Grid@Map[plotTrajectory3D[#, ImageSize → 330] &,
           Map[StringTemplate["He is from the city of `` , he speaks the language of"],
            {{"42", "carrot"}, {"chair", "doOr"}}, {2}], {2}]
```

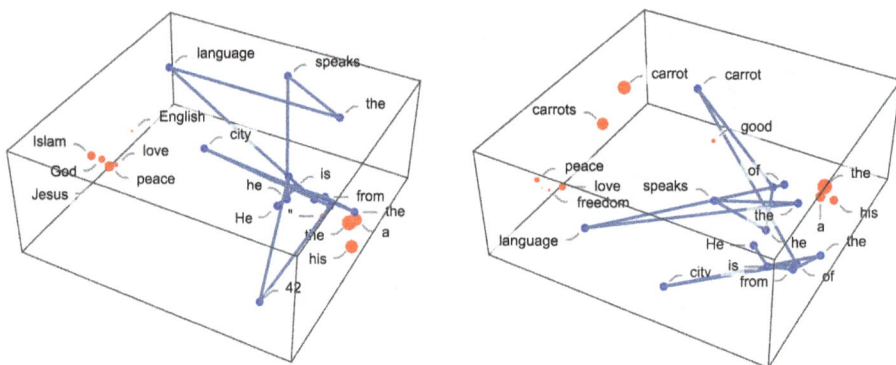

Out[ ]=

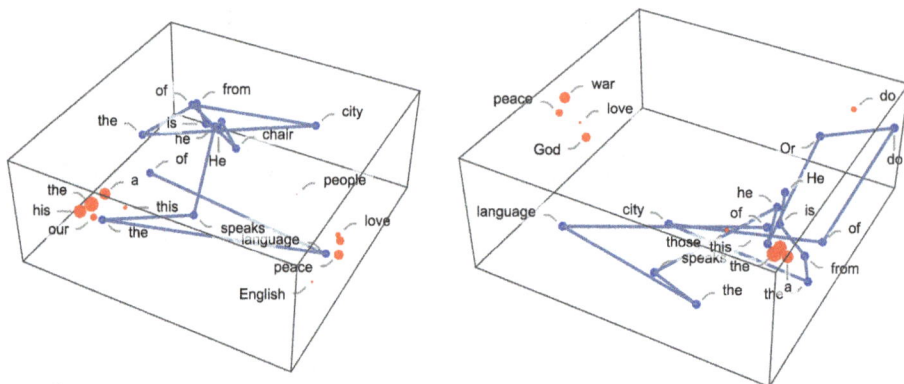

In[ ]:= `Grid@Map[plotTrajectoryGPT353D[♯, ImageSize → 330] &,`
`Map[StringTemplate["He is from the city of `` , he speaks the language of"],`
`{{"42", "carrot"}, {"chair", "doOr"}}, {2}], {2}]`

Out[ ]=

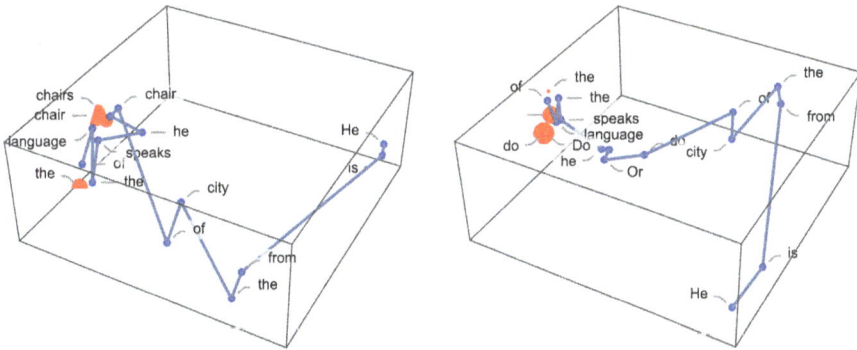

The contrast is most apparent when we put GPT-2 and GPT-3 side by side:

In[ ]:= `Row[{plotTrajectory3D[`
`StringTemplate["He is from the city of `` , he speaks the language of"][`
`"cabbage"], ImageSize → 330], plotTrajectoryGPT353D[`
`StringTemplate["He is from the city of `` , he speaks the language of"][`
`"cabbage"], ImageSize → 330]}]`

Out[ ]=

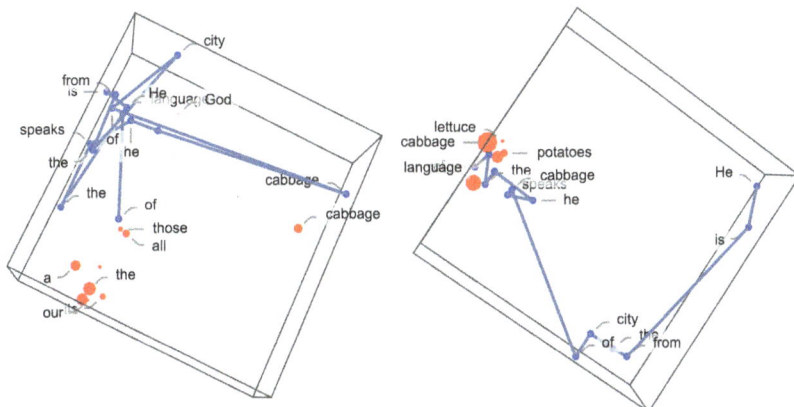

Concluding Remarks

LLMs are becoming critical tools for improving general human productivity, yet our understanding of the internal workings of these models remains limited. From GPT-2 to GPT-3 and GPT-4, intelligent behavior seemingly arises from the pure scaling of the model size, training data and slight modification of the tokenization scheme.

In this project, we compared the GPT-2 and GPT-3 models to investigate their trajectories in the embedding space. We found that GPT-3 is much more stable under small perturbations. A possible cause for such stability might come from its capability for contextualization.

Some potential future extensions of this work include the following:

- Our works are mostly qualitative. To concretize our claim, more quantitative study on the embedding trajectories is needed. It is possible to leverage methods from numerical analysis along with some existing mathematical formulations of transformer architecture to gain insights into the nature of the decision-making processes of LLMs.

- Parts of our study are not applicable due to the closed-source nature of the GPT-3.5 and GPT-4 models. If, in the future, OpenAI decides to make them open source, it should be fairly easy to apply the same workflow to them.

- To construct the visualization in 2D and 3D, we used the DimensionReduce function, which essentially applies principle component analysis (PCA) to the input $n \times 768$ matrix. A possible extension is to extract the variance explained by the dimension reduction method and perform different dimension reduction techniques.

- In this project, we chose to focus on a limited selection of sentences to reduce the space of inputs. However, in principle, the work can be easily extended to a vast array of sentences. Examples include prompts submitted to the Wolfram Prompt Repository.

- In the interpolation section, we chose to interpolate a pair of single tokens. A possible extension is to interpolate between n singular tokens.

In conclusion, there are many more works to be done in understanding LLMs. By computationally evaluating critical structures in LLMs, we hope to promote further study in the science of LLMs and ultimately harness their power to advance various domains and industries.

Acknowledgments

I would like to thank Bob Nachbar, Christopher Wolfram, Jofre Espigulé Pons and Mark Greenberg for their invaluable guidance in the overall development of the project. I am grateful to Stephen Wolfram for providing the initial idea that inspired this project. Additionally, I want to extend a special thanks to my mentor Alejandra Ortiz for carrying me throughout the project (and possibly burning down her laptop for running some heavy-lifting computations).

Reference

1. A. Radford, et al. (2019), "Language Models Are Unsupervised Multitask Learners." d4mucfpksywv.cloudfront.net/better-language-models/language-models.pdf.

Access the Full Code

Scan or visit wolfr.am/WSS2023-Luhan.

Additional Examples

Trajectories in the Embedding Space

```
In[•]:= Grid@Map[plotTrajectory3D, Map[StringTemplate["The capital of `` is"],
          {{"Japan", "Sweden"}, {"Australia", "China"}}, {2}], {2}]
```

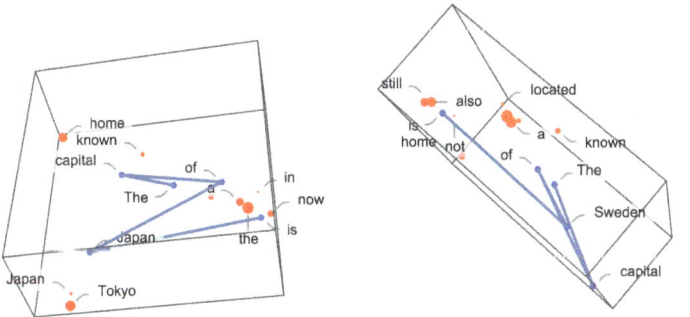

```
Out[•]=
```

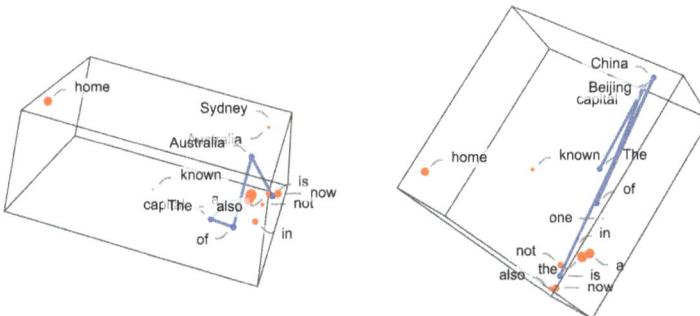

In[ ]:= Grid@Map[plotTrajectory3D, Map[StringTemplate["The `` is sleeping on the"],
{{"bear", "fox"}, {"man", "cat"}}, {2}], {2}]

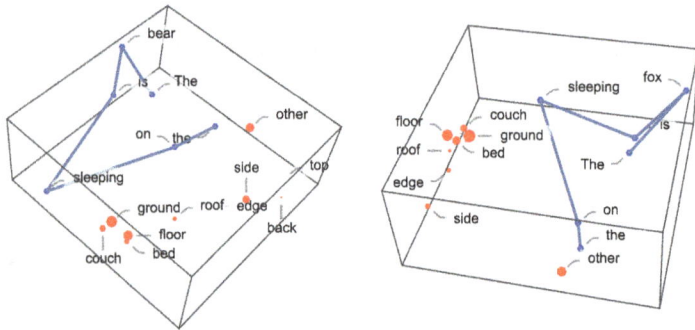

Out[ ]=

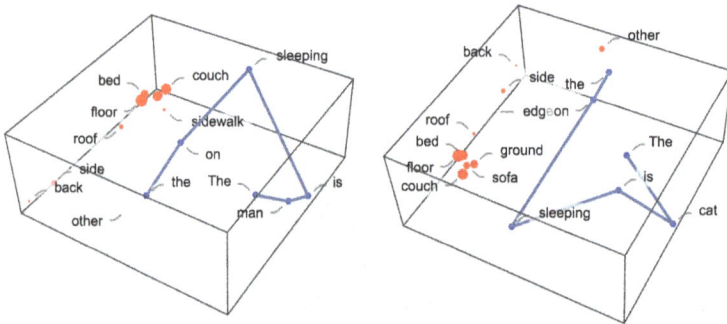

Top *n* Probabilities Predicted by the Model

In[ ]:= plotTopN[{"The man speaks English, he is from the city of",
"The man speaks French, he is from the city of",
"The man speaks Spanish, he is from the city of"}, 10]

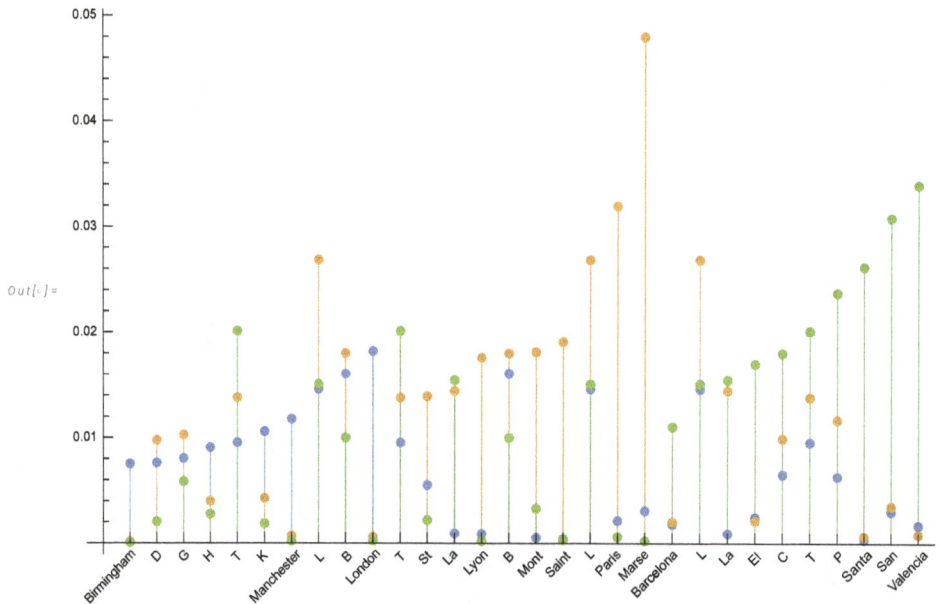

Out[ ]=

Similarity Matrix for Concept Space Mapping

In[ ]:= **interpolationTrajectory[" French",**
{" English", " Spanish", " scared", " math"}, Range[0, 1, 0.1]]

Out[ ]=

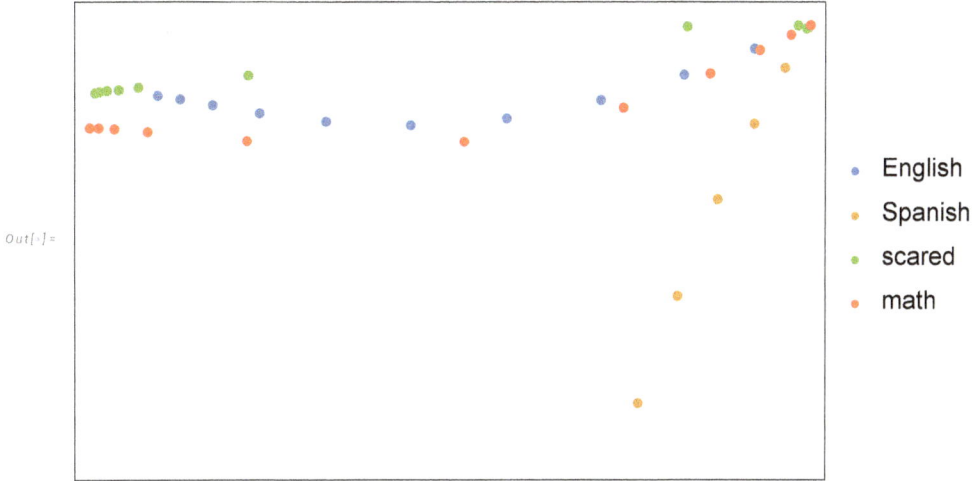

In[ ]:= **With[{sentence = "The cat sit on the"},**
Row[{subsentenceSimilarity[sentence, PlotLegends → "GPT2"],
subsentenceSimilarityGPT35[sentence, PlotLegends → "GPT3"]}]]

Out[ ]=

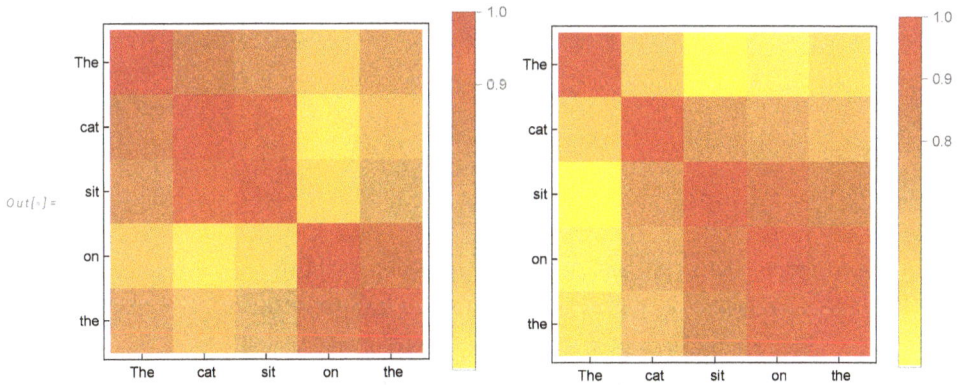

In[○]:= **With[{sentence = "After eating the cake, the plate was"},**
　　Row[{subsentenceSimilarity[sentence, PlotLegends → "GPT2"],
　　　subsentenceSimilarityGPT35[sentence, PlotLegends → "GPT3"]}]]

Out[○]=

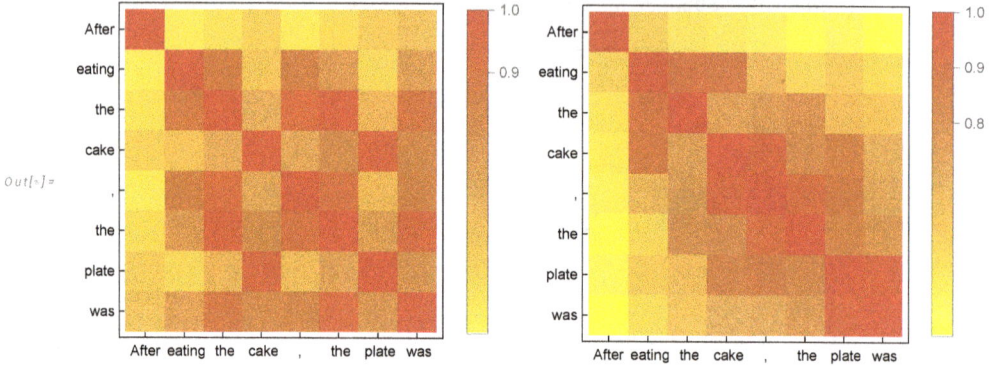

Out[○]= 13.3.0 for Mac OS X ARM (64–bit) (June 8, 2023)

In[○]:= **interpolationDivergence[" sat", {" stood", " sleep", " fly", " cook"},**
　　Range[0, 1, 0.1], StringTemplate["The cat`` on the"]]

Out[○]=

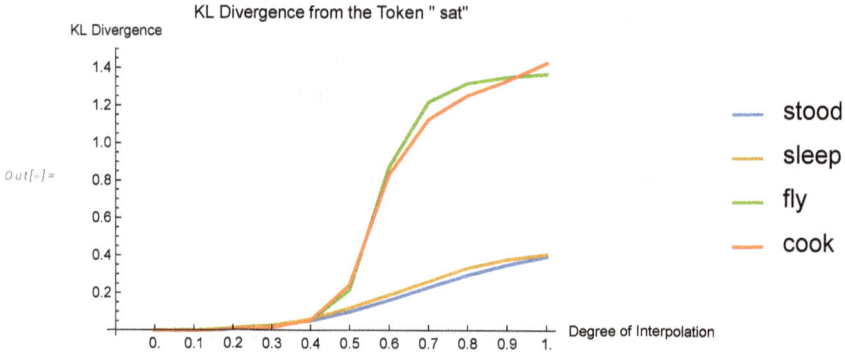

In[○]:= **Row[{plotSimilarity[Range[0, 1, 0.1], interpToken[**
　　　Range[0, 1, 0.1], {" sit", " cook"}, StringTemplate["The cat`` on the"]],
　　　PlotLabel → "Interpolation between token sat and cook"],
　　plotSimilarity[Range[0, 1, 0.1], interpToken[Range[0, 1, 0.1],
　　　{" sit", " stand"}, StringTemplate["The cat`` on the"]],
　　　PlotLabel → "Interpolation between token sat and sleep"]}]

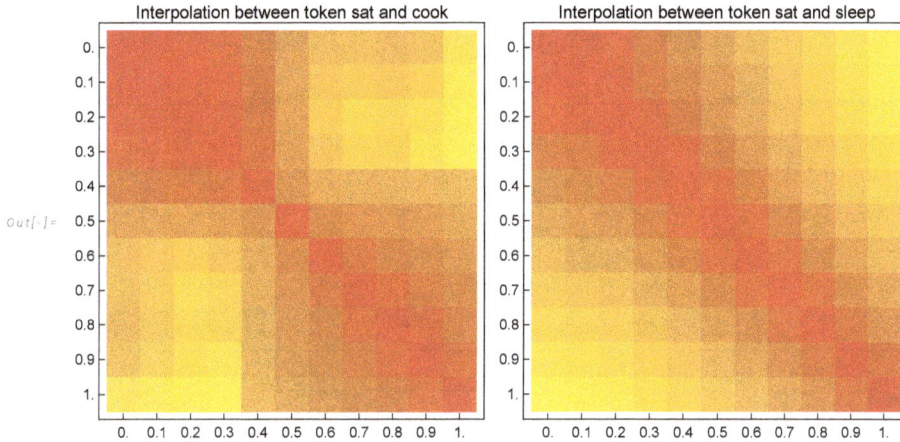

Interpolation between token sat and cook | Interpolation between token sat and sleep

Comparing Different GPT-2 Models

```
In[ ]:= gpt2lm117m = NetModel[{"GPT-2 Transformer Trained on WebText Data",
            "Task" → "LanguageModeling", "Size" → "117M"}];
       gpt2lm345m = NetModel[{"GPT2 Transformer Trained on WebText Data",
            "Task" → "LanguageModeling", "Size" → "345M"}];
       gpt2lm774m = NetModel[{"GPT2 Transformer Trained on WebText Data",
            "Task" → "LanguageModeling", "Size" → "774M"}];
```

```
In[ ]:= ClearAll[tokenize, vocabulary, foldTokens];
       tokenize = Information[gpt2lm117m, "InputPorts"]["Input"];
       vocabulary =
           Part[Last @ Last @ Normal @ Information[gpt2lm117m, "OutputPorts"], "Labels"];
       foldTokens[inputString_String] :=
         Part[vocabulary, #] & /@ FoldList[#1 ~ Join ~ {#2} &, {First @ tokenize
           @ inputString}, Rest @ tokenize @ inputString];
```

The embedding trajectory using a larger GPT-2 model:

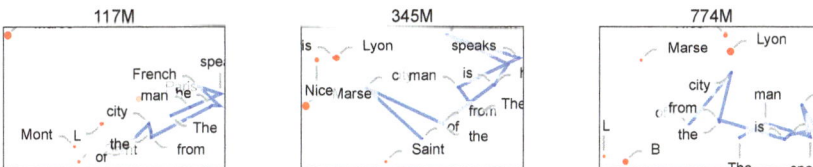

117M | 345M | 774M

Comparing GPT-3 Models from Different Families

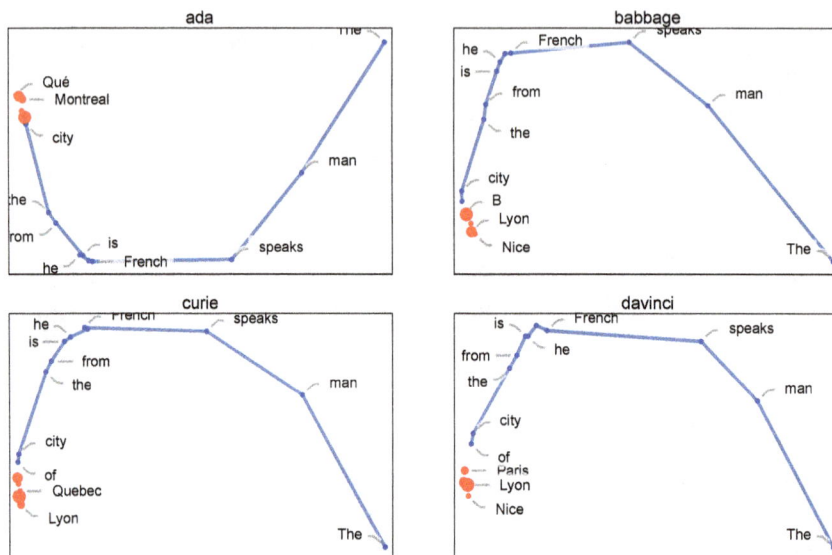

Cite This Notebook

"Study the Effect of Small Changes in Large Language Models"

by Luhan Cheng

Wolfram Community, STAFF PICKS, July 12, 2023

community.wolfram.com/groups/-/m/t/2958873

Animating Wolfram Model Evolutions in 3D

DUGAN HAMMOCK

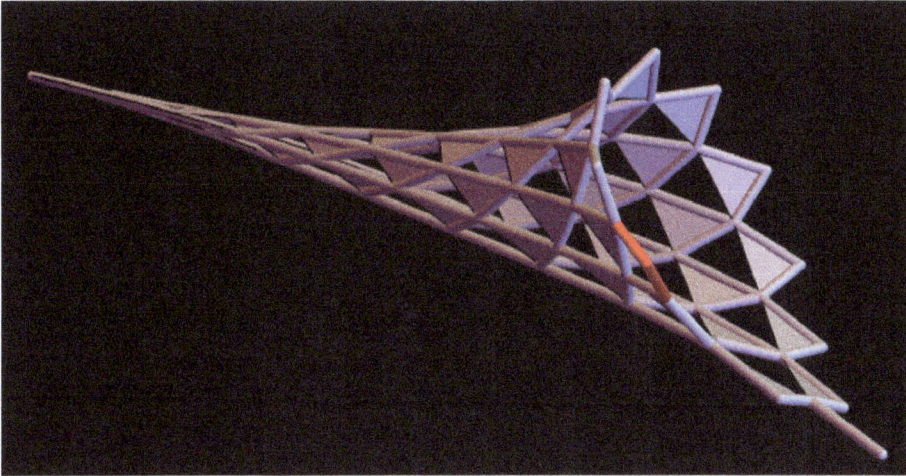

The evolution of hypergraphs undergoing successive rewriting events is a central theme for the Wolfram Physics Project. Individual hypergraphs are typically plotted using the GraphLayout *option* "SpringElectricalEmbedding", *which preserves and displays the overall shape and topology of the hypergraph. Animations of Wolfram models undergoing successive updating events have historically suffered from instability in the placement of vertices from one frame to the next. In this project, an algorithm is implemented that solves this problem and computes vertex coordinates for hypergraph animations in a temporally coherent manner.*

Access the Full Code

Scan or visit wolfr.am/WSS2023-Hammock.

The Problem: Incoherent Animations

The standard method for plotting hypergraphs undergoing sequential updating events is to use the "EventsStatesPlotsList" command for a Wolfram model:

```
In[ ]:=  wmEvolution =
           WolframModel[
             {{{1, 1, 2}, {1, 3, 4}} → {{4, 4, 3}, {2, 5, 3}, {2, 5, 3}}},
             {{1, 1, 1}, {1, 1, 1}},
             <|"MaxEvents" → 75|>
           ];
```

Each state in the evolution is plotted independently. No information from previous or subsequent states is used in the plot of an individual state. As a consequence, the placement of vertices and the overall orientation of the hypergraph may vary a lot from one state to the next.

In this example, the hypergraph swaps between left and right orientations:

```
In[ ]:=  wmStatePlots = wmEvolution["EventsStatesPlotsList"];
         Grid[Partition[Take[wmStatePlots, −12], 3]]
```

Out[ ]=

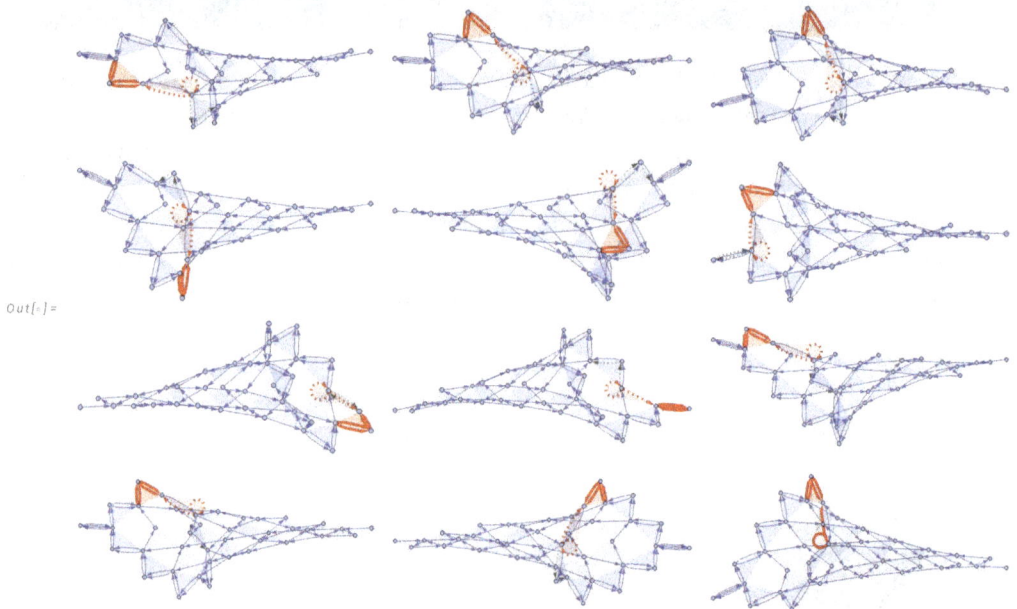

The Solution: Solving for VertexCoordinates

Stacking the Evolution and Adding Edges, Constraints

At the heart of the problem of incoherent animations is that each state in the evolution of the hypergraph is plotted independently, with no information being shared between states. Each vertex is free to be moved anywhere because it is not told to stay close to its location in the previous state.

A solution to this is to first compute the complete sequence of hypergraphs and assimilate them into a new stacked graph with edges added for vertices that persist from one event to the next.

This new graph includes the complete history of the evolution. The vertices of each event are confined to a separate layer, and interstitial edges are created between layers for vertices that survive from one event to the next. These added edges are what ensure temporal coherence for vertex positions in the animation.

Coordinate constraints are added using the VertexCoordinates option for GraphEmbedding, which inform the solver to keep each event confined to a separate layer:

In[ ]:= **wmEvolution // WolframModelAnimation2DStackPlot**

Out[ ]=

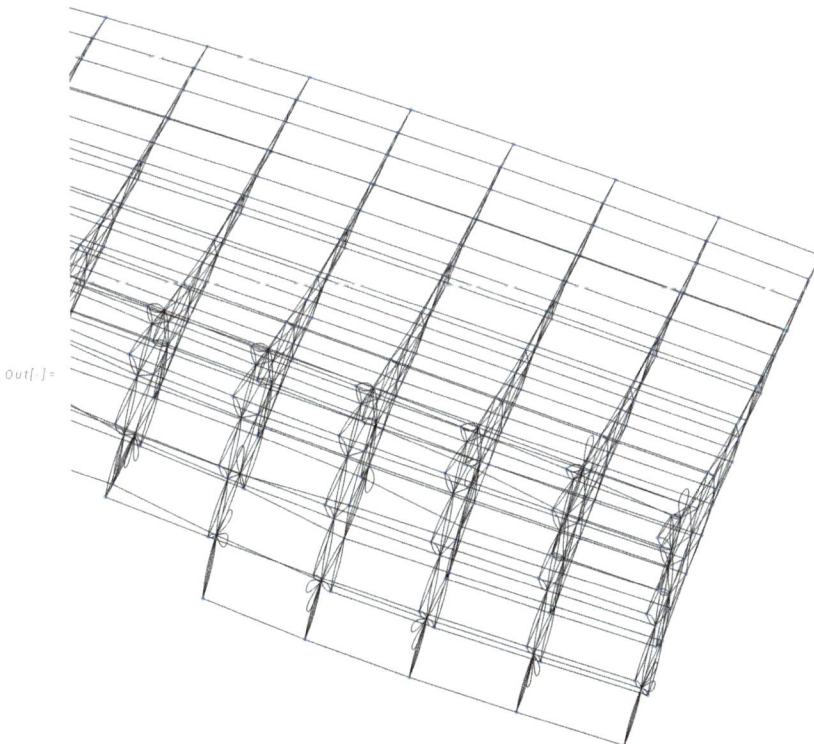

Extracting the Animation from the History Graph

This new graph, which incorporates the entire history, is fed into GraphEmbedding, along with the added interstitial edges and added layer constraints. The solver uses "SpringElectricalEmbedding" to compute vertex coordinates, and the derived animation is constructed by retrieving the graphs at each layer in sequence:

In[ ]:= **wmEvolution // WolframModelAnimation2D**

Out[ ]=

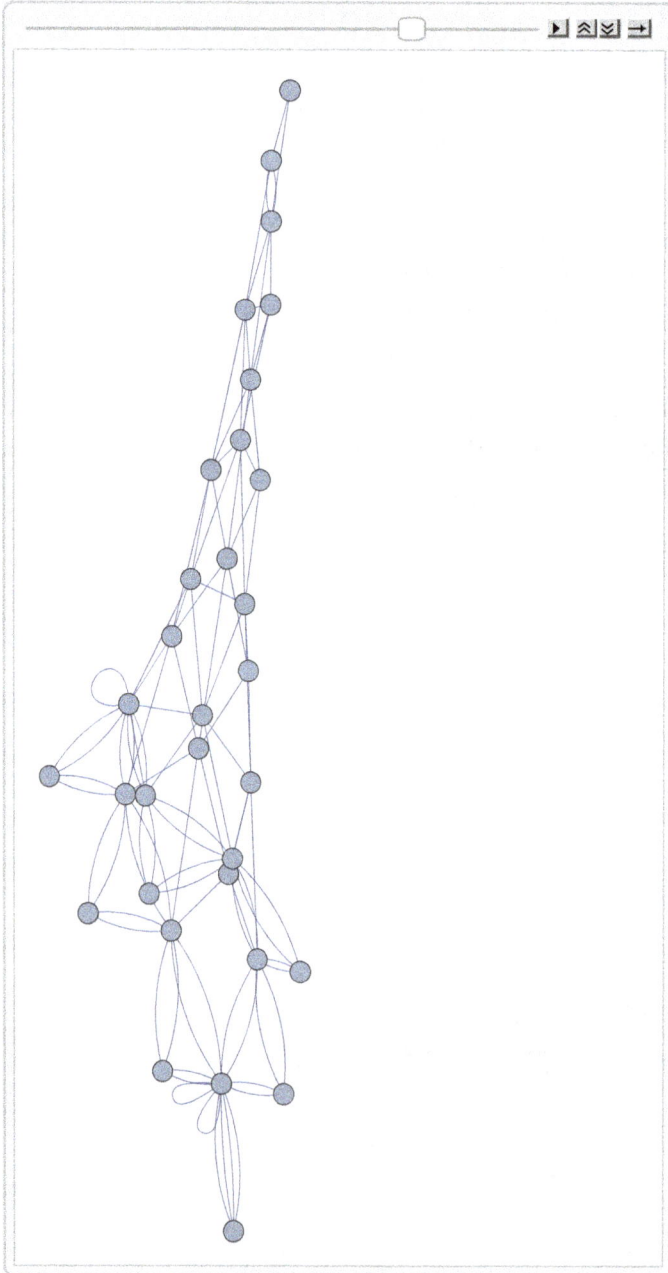

3D Animation of Wolfram Model

The algorithm used by GraphEmbedding is robust and general enough to solve for vertex coordinates in any dimension.

In order to make a 3D animation of an evolving Wolfram model, 3D graphs are stacked as layers into 4D space; interstitial edges are made for persistent vertices; coordinate constraints are added to keep the 3D graphs confined to their respective layers; and finally, the sequence of 3D hypergraphs is extracted from the 4D data.

The 3D hypergraphs are rendered using Line and Polygon graphics primitives. Highlighting has been added to show new (red) and deleted (green) hyperedges:

```
In[ ]:= wmEvolutionGraphicsComplexes3D = wmEvolution // WolframModelAnimation3D;

Manipulate[
  Graphics3D[
    wmEvolutionGraphicsComplexes3D[[eventIndex]]
    , Boxed → False
    , Axes → False
    , ImageSize → {400, 400}
    , Background → Black
  ]
  , {eventIndex, 1, Length[wmEvolutionGraphicsComplexes3D], 1}
  , SaveDefinitions → True
]
```

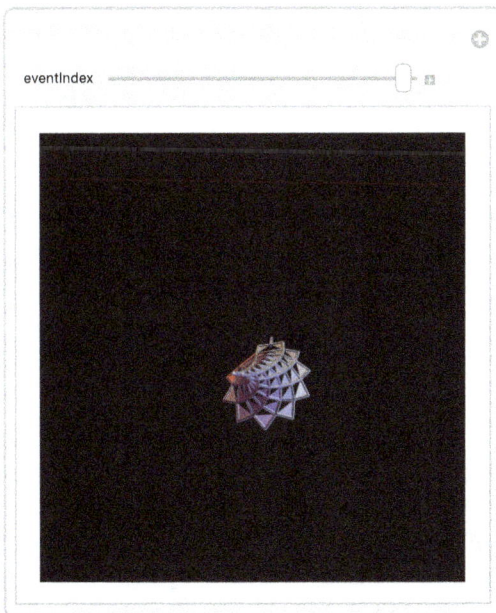

Examples of 3D Animations

Flower

```
In[*]:=  wmEvolution1 =
          WolframModel[
             {{{1, 1, 2}, {3, 1, 4}} → {{5, 5, 4}, {3, 4, 3}, {2, 1, 5}}},
             {{1, 1, 1}, {1, 1, 1}},
             <|"MaxEvents" → 100|>
          ];


        wmEvolution1GraphicsComplexes3D = wmEvolution1 // WolframModelAnimation3D;


        Manipulate[
          Graphics3D[
            wmEvolution1GraphicsComplexes3D[[eventIndex]]
            , Boxed → False
            , Axes → False
            , ImageSize → {400, 400}
            , Background → Black
          ]
          , {eventIndex, 1, Length[wmEvolution1GraphicsComplexes3D], 1}
          , SaveDefinitions → True
        ]
```

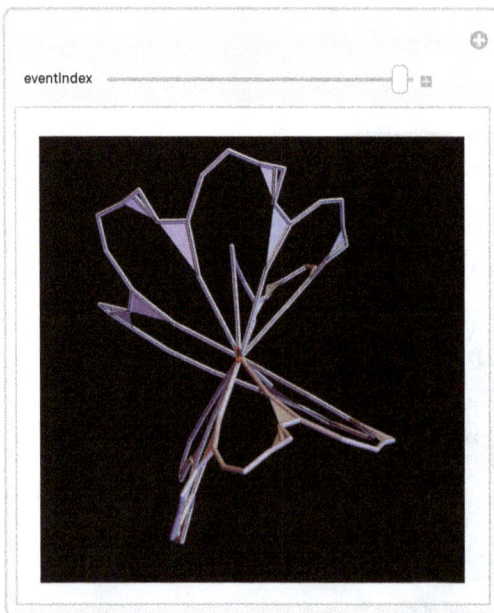

```
In[ ]:=  wmEvolution2 =
           WolframModel[
             {{{1, 2}, {2, 3}} → {{2, 4}, {2, 4}, {4, 1}, {3, 4}}},
             {{1, 1}, {1, 1}},
             <| "MaxEvents" → 50 |>
           ];
```

```
         wmEvolution2GraphicsComplexes3D = wmEvolution2 // WolframModelAnimation3D;
```

```
         Manipulate[
           Graphics3D[
             wmEvolution2GraphicsComplexes3D[[eventIndex]]
             , Boxed → False
             , Axes → False
             , ImageSize → {400, 400}
             , Background → Black
           ]
           , {eventIndex, 1, Length[wmEvolution2GraphicsComplexes3D], 1}
           , SaveDefinitions → True
         ]
```

Out[ ]=

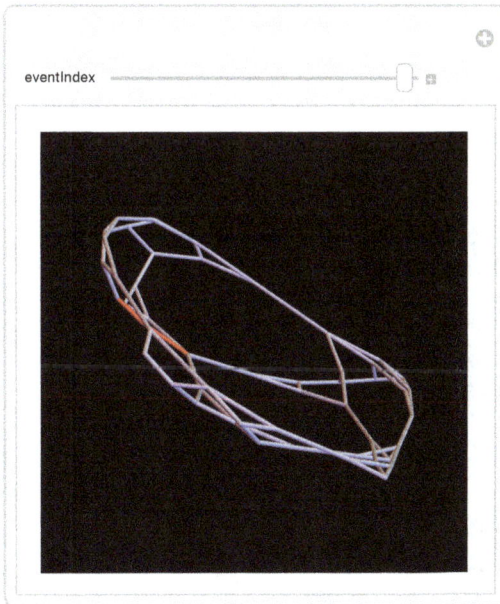

```
In[·]:=  wmEvolution3 =
            WolframModel[
              {{{1, 2}} → {{1, 2}, {2, 3}}},
              {{1, 1}},
              <| "MaxEvents" → 50 |>
            ];

         wmEvolution3GraphicsComplexes3D = wmEvolution3 // WolframModelAnimation3D;

         Manipulate[
           Graphics3D[
             wmEvolution3GraphicsComplexes3D[[eventIndex]]
             , Boxed → False
             , Axes → False
             , ImageSize → {400, 400}
             , Background → Black
           ]
           , {eventIndex, 1, Length[wmEvolution3GraphicsComplexes3D], 1}
           , SaveDefinitions → True
         ]
```

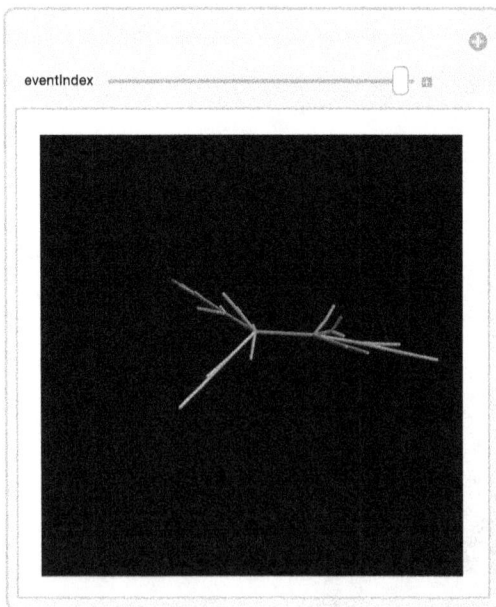

Conclusions and Future Work

This project presents a proof of concept for producing temporally coherent animations of hypergraph evolutions in three dimensions. There is a lot more development that can expand on this result:

- Adding options for customized styles and highlighting.
- Sometimes the animations have a long-range drift, which might be mitigated by adding more constraints for the GraphEmbedding solver:
  - Adding more interstitial edges or strengthening their spring constants in the solver could help.
  - Another solution might be to increase the mass of older vertices, making them harder to move compared to newly created vertices.
- Solving for the whole history at once is very cumbersome and animating hundred of states is infeasible:
  - Implementing a revised algorithm that solves for a few events at a time and stitching the results together would reduce computation time considerably.

Acknowledgments

Special thanks to Hatem Elshatlawy and Nik Murzin for their guidance and support during the Wolfram Summer School. I am also grateful for Max Piskunov for the work he has done with SetReplace and its helpful documentation. A personal thank-you goes out to Stephen Wolfram for his excitement and infectious enthusiasm, and to all instructors and TAs who helped make the Wolfram Summer School possible.

References

1. maxitg, SetReplace [Mathematica package]. *GitHub*. github.com/maxitg/SetReplace.

2. S. Wolfram (2020), "The Wolfram Physics Project," *The Wolfram Physics Project*. www.wolframphysics.org.

Cite This Notebook

"Animating Wolfram Model Evolutions in 3D"
by Dugan Hammock
Wolfram Community, STAFF PICKS, July 12, 2023
community.wolfram.com/groups/-/m/t/2959098

Hat Tiling Space Reduction and Grow Function Implementation

BOWEN PING

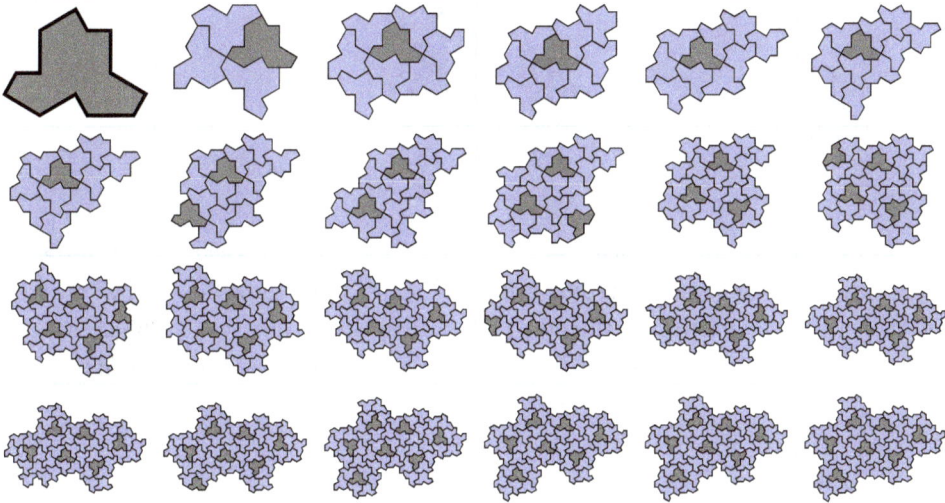

This project aims to implement a function to grow a given hat tile cluster automatically. At first, some reduction about specific vertex configurations are deducted to filter possible cases. Then a multiway grow function is implemented to study some special hat tile clusters. A dynamic module showing the grow process of clustering is implemented to illustrate the procedure with clarity. Additionally, a function with a fast algorithm to find all special subclusters in a big super-tile is obtained. The next step is to use more local vertex information to improve the growth algorithm, which will be useful in studying the hat tile space.

Hat Tiling

The hat tile is an aperiodic monotile, which can fill the whole infinite two-dimensional plane without translation symmetry. It looks like a "hat" or a "T-shirt":

```
In[·]:= GraphicsRow[{show[{{0, 0}, 0, False, 0}, ImageSize → Small],
        show[{{0, 0}, 0, True, 0}, ImageSize → Small], show[ {···} + ]}]
```

Out[·]=

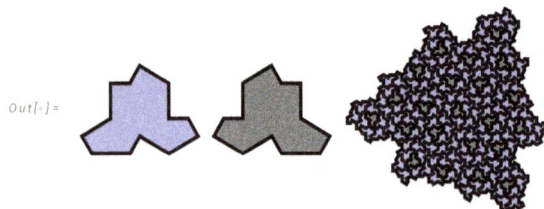

The left two figures are single hat tiles with different parity and the right one is a cluster, which can cover the whole plane after using substitution rules. Here, however, we are trying to implement a grow function rather than substitution rule to generate a hat tiling cluster. The first step is to study the vertex configurations of hat tiles and discover how they fit together.

Vertex Configuration Study

In order to implement a grow function to generate valid and promising tiling patterns, it's important to reduce as many invalid vertex configurations as possible. There are four different angles of a hat tile; if we take 2π as unit, then the angles are 1/4, 1/3, 2/3 and 3/4. Starting from the bottom concave vertex, we number each vertex from 0 to 12. There are several configurations for hats to meet and fill a point on the plane: 1/4 + 1/4 + 1/4 + 1/4, 1/3 + 1/3 + 1/3, 1/3 + 2/3 and 1/4 + 3/4. Additionally, there is a notable 1/2 + 1/4 + 1/4 vertex configuration with a characteristic "T" structure if we regard an edge as a 1/2 vertex. The 1/3 vertices combination has been studied by Brad Klee: community.wolfram.com/groups /-/m/t/2935078. We use black and brown to represent reflected hats, and yellow and blue for unreflected hats.

First, some preparations are needed for the convenience of future study.

Access the Full Code

Scan or visit wolfr.am/WSS2023-Ping.

Preparation Analysis

Intersection Test

First, an efficient intersection test method is needed to create a table recording all intersections of two hats sharing a vertex. I used the internal line test and area test to build the table twice and then choose the algorithm and code by Brad Klee to double-check and improve the table:

```
In[ ]:=  fixedHat[n_] := {{0, 0}, 0, False, n};
         moveHat[rotation_, ref_, pt_] := {{0, 0}, rotation, ref, pt};
```

Consider two hats: one is fixed, and the other can move around the first at different vertices, angles and parities. We can translate one hat along the vertex of the other hat, but its rotations and parity are fixed. We also change the state of the other one. What matters is the rotation angle difference and parity difference between these two hats; many cases can be reduced. Let them meet at point (0, 0). An intersection test function is also needed:

```
In[ ]:=  planarTest[hat1_, hat2_] := With[
             {
                 res = Length /@ PlanarPolygonFragmentation @@@ (hat @@@ {hat1, hat2})
             },
             If[hat1 == hat2, False, res == <| |> || Total[res] == 3]
         ]

In[ ]:=  testForTable[n_, rotation_, ref_, m_] :=
             planarTest[fixedHat[n], moveHat[rotation, ref, m]]
```

All possible cases of two hats are as follows:

```
In[ ]:=  GraphLayout
```

```
In[ ]:=  allcasesForIntersectionTable =
             Tuples[{Range[0, 12], Range[0, 11], {False, True}, Range[0, 12]}];
         allcasesForIntersectionTable // Length
```

```
Out[ ]=  4056
```

However, there are some way to reduce these cases. By regarding the state of (n, rotation, ref, m) as a dual state of (m, $r/-r$, ref, n) where if ref == False we take r; otherwise, we take $-r$ because of the reflection. So we can reduce almost half of the cases:

```
In[ ]:=  allcasesForIntersectionTable =
             Outer[{#1[[1]], #2[[1]], #2[[2]], #1[[2]]} &, Flatten[Table[{j, i},
             {i, 0, 12}, {j, 0, i}], 1], Tuples[{Range[0, 11], {False, True}}], 1];
         allcasesForIntersectionTable = SortBy[Flatten[allcasesForIntersectionTable, 1],
             (((((#[[1]] * 13) + #[[2]]) * 2) + Boole[#[[3]]]) * 13 + #[[4]]) &];
```

```
In[ ]:=  AbsoluteTiming[IntersectionTable = testForTable @@@ allcasesForIntersectionTable;]
```
```
Out[ ]=  {61.5302, Null}
```

Then we have a table recording all intersection cases. When we have two hats sharing same vertex as pt, $\{pt, n, dir_1, p_1\}$ and $\{pt, m, dir_2, p_2\}$, we have to transform their codes into the format shown previously and look up the table:

```
In[ ]:=  IntersectionTable = {...} + ;
         getIndexforIT = First[First[Position[allcasesForIntersectionTable, {###}]]] &;
         checkIntersectionTable[n_, r_, ref_, m_] :=
           IntersectionTable[[getIndexforIT[n, r, ref, m]]] /; n ≤ m
         checkIntersectionTable[n_, r_, ref_, m_] :=
           IntersectionTable[[getIndexforIT[m, If[ref, r, Mod[−r, 12]], ref, n]]] /; n > m
```

Since the two hats we check must share the same vertex in their code, we need to check and transform them:

```
In[ ]:=  findEqualHat[pos_, hatstate_] := With[
           {
                p = Simplify[pos]
           },
                With[{position = Position[Simplify[hatVertices @@ hatstate], p]},
                If[
                     position === {},
                     hatstate,
                {p, hatstate[[2]], hatstate[[3]], First[First[position]] − 1}]
                   ]
           ]
```

Then, by the previous analysis, we transform the code and check the table:

```
In[ ]:=  checkIntersectionofTwoHats[reference_, test_] := Module[
           {
                relative = findEqualHat[reference[[1]], test]
           },
                If[relative === reference, Return[False]];
                checkIntersectionTable[reference[[4]],
                     If[
                          Xor[relative[[3]], reference[[3]]],
                          Mod[relative[[2]] + reference[[2]], 12],
                          Mod[relative[[2]] − reference[[2]], 12]],
                     Xor[relative[[3]], reference[[3]]],
                     relative[[4]]]
           ]
```

Then we can visualize the result. The following is part of the table:

```
In[ ]:=  Grid[
            Partition[
              Labeled[
                  show[{fixedHat[#1], moveHat[#2, #3, #4]}, ImageSize → {100, 100}],
                  Style[checkIntersectionofTwoHats[
                      {{0, 0}, 0, False, #1}, {{0, 0}, #2, #3, #4}], Bold, Italic, Purple],
                  Bottom, ContentSize → {100, 100}] & @@@
               Tuples[{{8, 10}, {0, 1, 2}, {False, True}, {8, 10}}],
               6],
            Frame → All, FrameStyle → LightGray]
```

In this table, only when two hats do not intersect and do not form a hole, the result is True.

Angle Analysis

Generating all cases then checking intersection to delete is inefficient. However, there is a way to reduce some cases at the beginning. This idea is from Brad Klee, who uses 12 points around a vertex of the hat and numbers each of them, and by which points are in the hat and which are not we can easily distinguish hats with different angles and states at a point.

The reason we do this is that there are only 12 directions for all edges of a hat:

```
In[ ]:=  allStateofOneHat =
            Prepend[#, {0, 0}] & /@ Tuples[{Range[0, 11], {False, True}, Range[0, 12]}];
```

```
In[ ]:=  DeleteDuplicates[(ArcTan @@ (#2 – #1) &) @@@
            (Union @@ (hatEdges @@@ allStateofOneHat))] // Length
```

```
Out[ ]=  12
```

Numbering each point from 1 to 12, it is easy to see the angle state at a certain point:

```
In[ ]:=  Show[
            show[{{0, 0}, 0, False, 0}],
            Graphics[Point[CirclePoints[{0, 0}, {1/4, Pi/12}, 12]]]
          ]
```

Out[ ]=

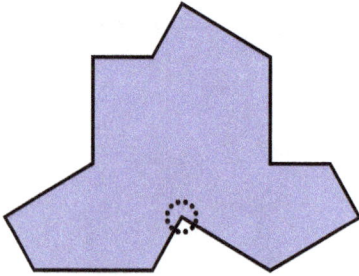

It is easy to get an association for each vertex and its point occupation:

```
In[ ]:=  vertexAngles = Association[
            {
                0 → 3/4,
                1 → 1/3,
                2 → 1/4,
                3 → 1/3,
                4 → 3/4,
                5 → 1/3,
                6 → 1/4,
                7 → 2/3,
                8 → 1/4,
                9 → 2/3,
                10 → 1/4,
                11 → 1/3,
                12 → 1/3
            }];
```

```
In[ ]:=  oneThirdVertices = Keys @ Select[vertexAngles, # == 1/3 &];
         oneFourthVertices = Keys @ Select[vertexAngles, # == 1/4 &];
         twoThirdVertices = Keys @ Select[vertexAngles, # == 2/3 &];
         threeFourthVertices = Keys @ Select[vertexAngles, # == 3/4 &];
```

```
In[ ]:=  pointToAngle = Association[{
                0 → {12, 1, 2, 3, 4, 5, 6, 7, 8},
                1 → {2, 3, 4, 5},
                2 → {5, 6, 7},
                3 → {7, 8, 9, 10},
                4 → {4, 5, 6, 7, 8, 9, 10, 11, 12},
                5 → {6, 7, 8, 9},
                6 → {9, 10, 11},
                7 → {7, 8, 9, 10, 11, 12, 1, 2},
                8 → {10, 11, 12},
                9 → {8, 9, 10, 11, 12, 1, 2, 3},
                10 → {11, 12, 1},
                11 → {1, 2, 3, 4},
                12 → {3, 4, 5, 6}
         }];
```

```
In[ ]:=  getPointsOccupation[hat_] :=
            Mod[(1 – 2 Boole[#2]) (pointToAngle[#3] + #1) + 7 (Boole[#2]),
         12, 1] & @@ hat /; Length[hat] == 3
         getPointsOccupation[hat_] :=
            Mod[(1 – 2 Boole[#2]) (pointToAngle[#3] + #1) + 7 (Boole[#2]),
         12, 1] & @@ hat[[2 ;;]] /; Length[hat] == 4
```

It's enough to analyze when some hats meet at one point at vertices, but here is a case when two hats meet at the midpoint of an edge of another hat:

```
In[ ]:=  Show[
            show[{{{3/2, Sqrt[3]/2}, 0, False, 0}, {{0, 0}, 0, False, 8}, {{0, 0}, 2, False, 2}}],
            Graphics[Point[CirclePoints[{0, 0}, {1/2, Pi/12}, 12]]]
         ]
```

Out[ ]=

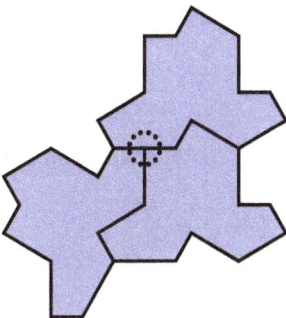

So it is necessary to calculate the angle occupations by edges in different directions. As we discussed before, there are only 12 directions for any edge. These are:

```
In[·]:= DeleteDuplicates[(ArcTan @@ (#2 - #1) &) @@@
          (Union @@ (hatEdges @@@ allStateofOneHat))] // Sort
```

$$
Out[·]= \left\{0, -\frac{5\pi}{6}, -\frac{2\pi}{3}, -\frac{\pi}{2}, -\frac{\pi}{3}, -\frac{\pi}{6}, \frac{\pi}{6}, \frac{\pi}{3}, \frac{\pi}{2}, \frac{2\pi}{3}, \frac{5\pi}{6}, \pi\right\}
$$

Then build the function for getting the angle configuration of edges:

```
In[·]:= vertexOnEdgeQ[vert_, edge_] := With[
        {
            vec1 = edge[[1]] – vert,
            vec2 = vert – edge[[2]]
            },
            vec1[[1]] == vec1[[2]] || vec2[[1]] == vec2[[2]] || And[ Dot[vec1, vec2] ≥ 0,
            ArcTan @@ vec1 === ArcTan @@ vec2]
        ]
```

```
In[·]:= edgeAngleToPoints = Association[
            Function[line,
                With[
                        {offset = Mod[
                                Floor[
                                        Simplify[
                                        (ArcTan @@ line + Pi / 12) / (2 Pi / 12)
                                        ]
                                ]
                        ], 12]
                },
                # → Sort[(Mod[#, 12, 1] &) /@ Range[1 + offset, 6 + offset]]
            ]][ReIm[Exp[I (#)]]]] & /@ {-((5 π) / 6), -((2 π) / 3), -(π / 2), -(π / 3),
            -(π / 6), 0, π / 6, π / 3, π / 2, (2 π) / 3, (5 π) / 6, π}];
```

```
In[·]:= getEdgeAngle = Function[{pt, edgeCoveringHats},
            Apply[
                If[
                        #2 == False,
                        Apply[ArcTan @@ (#2 – #1) &, #1, {1}],
                        Apply[ArcTan @@ (#1 – #2) &, #1, {1}]
                ] &,
                Function[hatstate,

                    {If[vertexOnEdgeQ[pt, #], #, Nothing] & /@ (hatEdges @@ hatstate),
                        hatstate[[3]]}
                    ] /@ edgeCoveringHats
            , {1}]
        ];
```

With this tool, many invalid cases would not be generated at first, which improves the efficiency significantly.

Obvious Invalid Cases

There are some obvious invalid cases, which were derived during the study. A summary table is helpful to filter possible cases:

```
In[ ]:=  knownInvalidCases = { ⋯ ⊕ };
```

```
In[ ]:=  testKnownInvalidCases = Function[onestate, With[
           {
              deleterule = Alternatives @@ (
                 Join @@ With[
                    {
                    seq =
                    ((

                                {OrderlessPatternSequence @@
                                    {___, ♯1, ___, ♯2, ___}}
                                & @@@ allRotationsofHats[♯] &
                                ) /@ knownInvalidCases)
                    },
                    Join[seq, seq /. {True → False, False → True}]
                    ]
                 )
              },
              MatchQ[onestate, deleterule]
              ]
           ];
```

Some of them are shown here:

```
In[ ]:=  GraphicsGrid[
           Partition[
              show /@ (Map[Prepend[♯, {0, 0}] &, knownInvalidCases, {2}][[ ;; 30]]),
              10],
           FrameStyle → LightGray, Frame → All, ImageSize → Full
           ]
```

1/4 + 3/4 Vertex Configuration

Get all 1/4 and 3/4 vertices, combine them and delete intersected and invalid cases. Only eight of them survive:

```
In[ ]:= threeFourthVertices = Select[pointToAngle, Length[#] == 9 &] // Keys
```
```
Out[ ]= {0, 4}
```

```
In[ ]:= oneFourthVertices = Select[pointToAngle, Length[#] == 3 &] // Keys
```
```
Out[ ]= {2, 6, 8, 10}
```

```
In[ ]:= buildState[pt1_, dir_, ref_, pt2_] := {{{0, 0}, 0, False, pt1}, {{0, 0}, dir, ref, pt2}};
```
```
In[ ]:= allcases = Tuples[{threeFourthVertices, Range[0, 11], {False, True}, oneFourthVertices}];
```
```
In[ ]:= noIntersectionCases = If[checkIntersectionTable[##], {##}, Nothing] & @@@ allcases;
```
```
In[ ]:= validCases = If[MemberQ[knownInvalidCases, #], Nothing, #] & /@
            (buildState @@@ noIntersectionCases)[[All, All, 2 ;;]];
```

```
In[ ]:= GraphicsRow[show /@ Map[Prepend[#, {0, 0}] &, validCases, {2}],
            Frame → All, FrameStyle → LightGray, ImageSize → Full]
```

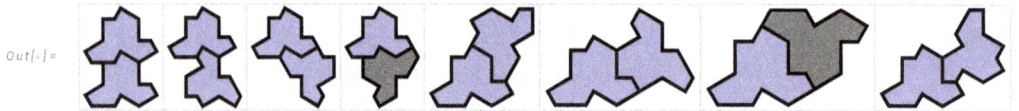

```
In[ ]:= validCases = Join[validCases, validCases[[{4, 7}]] /. {True → False, False → True}];
        GraphicsRow[show /@ Map[Prepend[#, {0, 0}] &, validCases, {2}],
            Frame → All, FrameStyle → LightGray, ImageSize → Full]
```

Consider reflection; the fourth and seventh can be reflected.

1/3 + 2/3 Vertex Configuration

As before, generate promising candidates and filter. Only 10 of them survive:

```
In[ ]:= oneThirdVertices = Select[pointToAngle, Length[#] == 4 &] // Keys
```
```
Out[ ]= {1, 3, 5, 11, 12}
```

```
In[ ]:= twoThirdVertices = Select[pointToAngle, Length[#] == 8 &] // Keys
```
```
Out[ ]= {7, 9}
```

```
In[ ]:= buildState[pt1_, dir_, ref_, pt2_] := {{{0, 0}, 0, False, pt1}, {{0, 0}, dir, ref, pt2}};
```

```
In[•]:=  allcases = Tuples[{twoThirdVertices, Range[0, 11], {False, True}, oneThirdVertices}];
```

```
In[•]:=  noIntersectionCases = If[checkIntersectionTable[##], {##}, Nothing] & @@@ allcases;
```

```
In[•]:=  validCases = If[MemberQ[knownInvalidCases, #], Nothing, #] & /@
             (buildState @@@ noIntersectionCases)[[All, All, 2 ;;]];
```

```
In[•]:=  GraphicsRow[show /@ Map[Prepend[#, {0, 0}] &, validCases, {2}],
             ImageSize → Full, FrameStyle → LightGray, Frame → All]
```

Out[•]=

1/4 + 1/4 + 1/4 + 1/4 Vertex Configuration

Get all 1/4 vertices and all states of one hat:

```
In[•]:=  oneFourthVertices = Keys @ Select[pointToAngle, Length[#] == 3 &]
```

```
Out[•]=  {2, 6, 8, 10}
```

```
In[•]:=  allStateforOnehat = Tuples[{Range[0, 11], {False, True}, oneFourthVertices}];
```

Group all states by their angle occupations:

```
In[•]:=  temp = GroupBy[allStateforOnehat, getPointsOccupation];
```

Create some temporary functions for the convenience of analysis:

```
In[•]:=  getAvailableCasesFunc = Function[{availablePoints, as},
             Block[{
                 res = {}
             },
             If[
                     SubsetQ[availablePoints, #],
                     res = Join[res, as[#]]
                 ] & /@ Keys[as]; res]
         ];
```

```
In[•]:=  getAvailableCasesListFunc = (
                 #
                   →
                 getAvailableCasesFunc[
                     Complement[Range[12], getPointsOccupation[#]], temp]
             ) &;
```

```
In[•]:=  tempCheckFunc[hat1_, hat2_] :=
             checkIntersectionofTwoHats @@ ({{0, 0}, Sequence @@ ##} & /@ {hat1, hat2})
```

Imagine a point waiting to be filled. Add hats one by one:

- First hat

First, create all possible cases where this is only one hat:

```
In[ ]:= res1 = Association[getAvailableCasesListFunc /@ allStateforOnehat];
       Total[Length /@ res1]
```

```
Out[ ]= 5376
```

Check the intersection table to filter cases:

```
In[ ]:= res2 = Association[
           Function[state, state → Select[res1[state], tempCheckFunc[state, #] &]] /@
               allStateforOnehat];
       Total[Length /@ res2]
```

```
Out[ ]= 3816
```

■ Second hat

Now add the second hat. Organize the previous result first:

```
In[ ]:= res3 = Union @@ ((Function[second, {#, second}] /@ res2[#]) & /@ Keys[res2]);
       res3 // Length
```

```
Out[ ]= 3816
```

```
In[ ]:= GraphicsRow[(show /@ Map[Prepend[#, {0, 0}] &, res3, {2}])[[Range[1, 3000, 300]]],
         ImageSize → Full, FrameStyle → LightGray, Frame → All]
```

```
Out[ ]=
```

Apparently there are many obvious invalid cases that need to be deleted. The least angle of a
hat is 1/4, which corresponds to three points. Thus, if the points left are divided into small,
continuous pieces—any of whose length is smaller than 3—they must be invalid. So I build a
function to cluster points from 1 to 12:

```
In[ ]:= clusterPoints = Function[list,
           Block[{p},
               If[
                      list == Range[12],
                      {Range[12]},
                      DeleteDuplicates[MapThread[Union, {Most[NestWhileList[
                      Mod[#+1,12,1]&,#,MemberQ[list,#]&]]&/@list,
                          Most[NestWhileList[Mod[#-1,12,1]&,#,MemberQ[list,
                          #]&]]&/@list}],
                      #1==#2&]
                  ]
               ]
           ];
```

Select those with valid continuous points:

```
In[ ]:= res4 = Select[res3, AllTrue[Length /@ clusterPoints[Complement[Range[12],
            Union @@ (getPointsOccupation /@ #)]], # ≥ 3 &] &];
      res4 // Length
```

```
Out[ ]= 1608
```

```
In[ ]:= GraphicsRow[(show /@ Map[Prepend[#, {0, 0}] &, res4, {2}])[[Range[1, 1608, 160]]],
      ImageSize → Full, FrameStyle → LightGray, Frame → All]
```

Organize the result again, and get its available angle points:

```
In[ ]:= res5 = (# → Intersection @@ Association[getAvailableCasesListFunc /@ #]) & /@ res4;
      res5 = Select[Association @ res5, # ≠ {} &];
```

■ Third hat

Now let's add the third hat:

```
In[ ]:= res6 = Union @@ ((Function[third, Append[#, third]] /@ res5[#]) & /@ Keys[res5]);
      res6 // Length
```

```
Out[ ]= 39 168
```

```
In[ ]:= GraphicsRow[show /@ RandomSample[((Map[Prepend[#, {0, 0}] &, res6, {2}])), 10],
      ImageSize → Full, Frame → All, FrameStyle → LightGray]
```

There are many intersected cases that need to be reduced, which takes a while:

```
In[ ]:= AbsoluteTiming[
        res7 = Select[res6, And @@ (tempCheckFunc @@@ Subsets[#, {2}]) &];]
      res7 // Length
```

```
Out[ ]= {89.1845, Null}
```

```
Out[ ]= 16 608
```

```
In[ ]:= GraphicsRow[show /@ RandomSample[((Map[Prepend[#, {0, 0}] &, res7, {2}])), 10],
      ImageSize → Full, Frame → All, FrameStyle → LightGray]
```

Now we can see that some points have been filled already, but are not really occupied by four 1/4 vertices. Some impossible gaps remain, and we cannot fill them with the hat tiles. So they need to be deleted:

```
In[ ]:= AbsoluteTiming[
    res8 = (♯ → Intersection @@ Association[getAvailableCasesListFunc /@ ♯]) & /@ res7;
    res8 = Select[Association[res8], Length[♯] ≠ 0 &];
    res8 // Length
    ]
```

```
Out[ ]= {31.175, 10 368}
```

■ Fourth hat

Now comes the last hat. Build all possible cases first:

```
In[ ]:= res9 = Union @@ ((Function[fourth, Append[♯, fourth]] /@ res8[♯]) & /@ Keys[res8]);
    res9 // Length
```

```
Out[ ]= 82 944
```

```
In[ ]:= GraphicsRow[show /@ RandomSample[((Map[Prepend[♯, {0, 0}] &, res9, {2}])), 10],
    ImageSize → Full, Frame → All, FrameStyle → LightGray]
```

```
In[ ]:= AbsoluteTiming[
    res10 = Select[res9, AllTrue[tempCheckFunc @@@ Subsets[♯, {2}], ♯ == True &] &];]
    res10 // Length
```

```
Out[ ]= {371.298, Null}
```

```
Out[ ]= 19 008
```

Since it takes six minutes to finish calculating, the result is iconized:

```
In[ ]:= res10 = {...} ⊹ ;
```

```
In[ ]:= GraphicsRow[show /@ RandomSample[((Map[Prepend[♯, {0, 0}] &, res10, {2}])), 10],
    ImageSize → Full, Frame → All, FrameStyle → LightGray]
```

Now we need to delete all known invalid cases, which takes a much longer time:

```
In[ ]:= AbsoluteTiming[
         res11 =
          Select[
            res10,
            !Or@@(
                   Function[invalid, Or[SubsetQ[#, invalid]]]/@
                     (Join@@allReflectionofHat[Join@@
                        (allRotationsofHats/@knownInvalidCases)]&[#])) &
             ];
          res11 // Length
          ]
Out[ ]= {1029.73, 2304}
```

The result is iconized:

```
In[ ]:= res11 = {...} + ;
```

```
In[ ]:= GraphicsRow[show/@RandomSample[((Map[Prepend[#, {0, 0}] &, res11, {2}])), 10],
          ImageSize → Full, Frame → All, FrameStyle → LightGray]
```

From the result, there are many cases that are reflected or rotated to each other.
Reduce them:

```
In[ ]:= AbsoluteTiming[res12 = DeleteDuplicates[
              res11,
              MemberQ[
                 Sort/@(Join@@allReflectionofHat[allRotationsofHats[#1]]), Sort[#2]] &
             ];
          res12 // Length
          ]
Out[ ]= {2.21855, 5}
```

Canonicalize the result and make unreflected hats the majority:

```
In[ ]:= res13 = If[Count[#, True, Infinity] > Count[#, False, Infinity],
           # /. {True → False, False → True}, #] & /@ res12;
```

```
In[ ]:= GraphicsRow[show /@ Map[Prepend[#, {0, 0}] &, res13, {2}],
         ImageSize → Full, Frame → All, FrameStyle → LightGray]
```

Out[ ]=

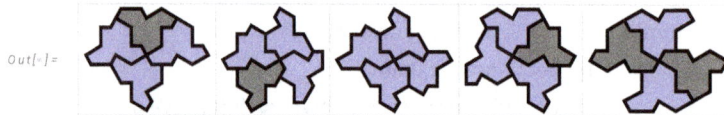

1/4 + 1/4 + 1/2 "T" Configuration

There is a special way for three hats to fill a point on the plane. Two hats meet at the middle point of another hat's longest edge, forming a "T"-shape configuration.

First, choose a hat as the "edge hat" and add two more to meet at the middle point of its edge:

```
In[ ]:= EdgeHat = {{0, 0}, 0, False, 0};
       meetPoint = Mean[(hatEdges @@ EdgeHat)[[-2]]];
```

```
In[ ]:= Show[
         show @ EdgeHat,

         Graphics[{Red, PointSize[.06], Point[{-3/2, -√3/2}]}],

         ImageSize → Small
       ]
```

Out[ ]=

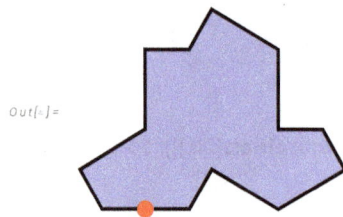

Here, we can build a simple grow function with the previous vertex configuration conclusions, which makes it easy and fast to get a few promising cases. However, implementing this kind of grow function will occupy too much space for new code, so I chose to use the intersection test from Brad Klee to filter cases, which is also fast enough:

```
In[ ]:= allStateforOnehat =
         Prepend[#, meetPoint] & /@ Tuples[{Range[0, 11], {False, True}, Range[0, 12]}];
       allStateforOnehat // Length
```

Out[ ]= 312

```
In[ ]:= AbsoluteTiming[res1 = Select[allStateforOnehat, FragmentationTest[{EdgeHat, #}] &];]
```

```
Out[ ]= {9.12001, Null}
```

```
In[ ]:= GraphicsRow[show[{#, EdgeHat}] & /@ RandomSample[res1, 10],
        ImageSize → Full, Frame → All, FrameStyle → LightGray]
```

There are many impossible cases, which can be seen just by looking at the results or found
by using the angle points method.

The fixed hat occupies half of the points of the upper half-plane. The available angle represented by points is a set of {7, 8, 9, 10, 11, 12}. Then get each point occupation of res1 to
generate corresponding possible cases:

```
In[ ]:= pts = Complement[Range[12],
            Union @@ (edgeAngleToPoints @@@ getEdgeAngle[meetPoint, {EdgeHat}])];
        pts
```

```
Out[ ]= {7, 8, 9, 10, 11, 12}
```

```
In[ ]:= res2 = getAvailableCasesFunc[Complement[pts, getPointsOccupation[#]], temp] & /@
            res1;
        res2 // Length
```

```
Out[ ]= 4 /
```

```
In[ ]:= res3 = Flatten[Function[i, {EdgeHat, res1[[i]], {meetPoint, ###}} & @@@ res2[[i]]] /@
            Range[Length[res1]], 1];
        res3 // Length
```

```
Out[ ]= 104
```

```
In[ ]:= GraphicsRow[show /@ RandomSample[res3, 10],
        ImageSize → Full, Frame → All, FrameStyle → LightGray]
```

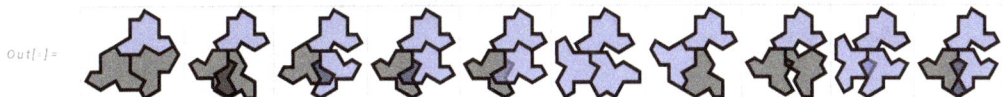

Use the intersection test to filter cases:

```
In[•]:= AbsoluteTiming[
          res4 = Select[res3, FragmentationTest[#] &];
          res4 // Length
        ]
```

```
Out[•]= {9.04139, 38}
```

Delete the same hat tile, but use a different order of codes:

```
In[•]:= res5 = DeleteDuplicatesBy[res4, Sort];
        res5 // Length
```

```
Out[•]= 19
```

```
In[•]:= GraphicsGrid[Partition[show /@ res5, UpTo[10]],
          ImageSize → Full, Frame → All, FrameStyle → LightGray]
```

Out[•]=

There are only 19 cases. Only cases 2, 3, 4 and 15 are promising:

```
In[•]:= res6 = res5[[{2, 3, 4, 15}]];
        GraphicsRow[show /@ res6, Frame → All, FrameStyle → LightGray]
```

Out[•]=

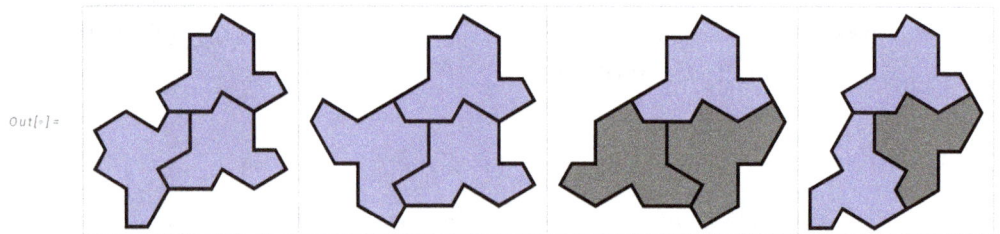

According to [1], the third one shows up as its reflected version. So there we change it into its reflected version too.

The result is:

```
In[•]:= TshapeList = { ... };
```

In[ ]:= **GraphicsRow[show /@ TshapeList, Frame → All, FrameStyle → LightGray]**

Out[ ]=

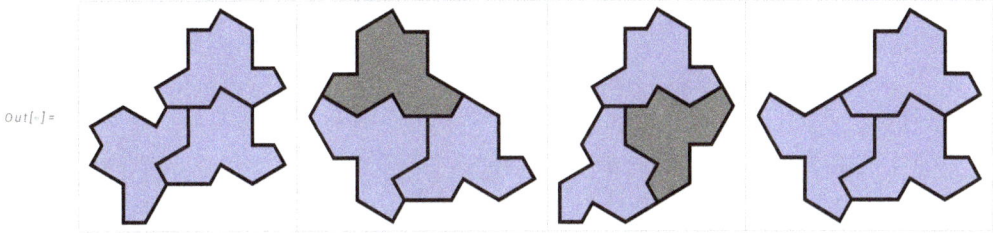

Conclusion

Vertex configurations summary:

In[ ]:= **oneThirdVertexTheoremLis={ ··· };**

oneFourthAndThreeFourthTheoremList={ ··· };

oneThirdAndTwoThirdTheoremList={ ··· };

oneFourthVertexTheoremLis = { ··· };

TshapeList={ ··· };

In[ ]:= Column[...]

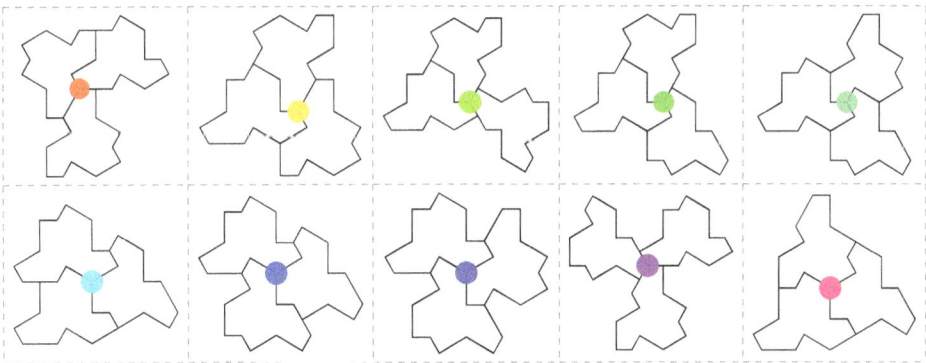

10 1/3 Vertex Configuration

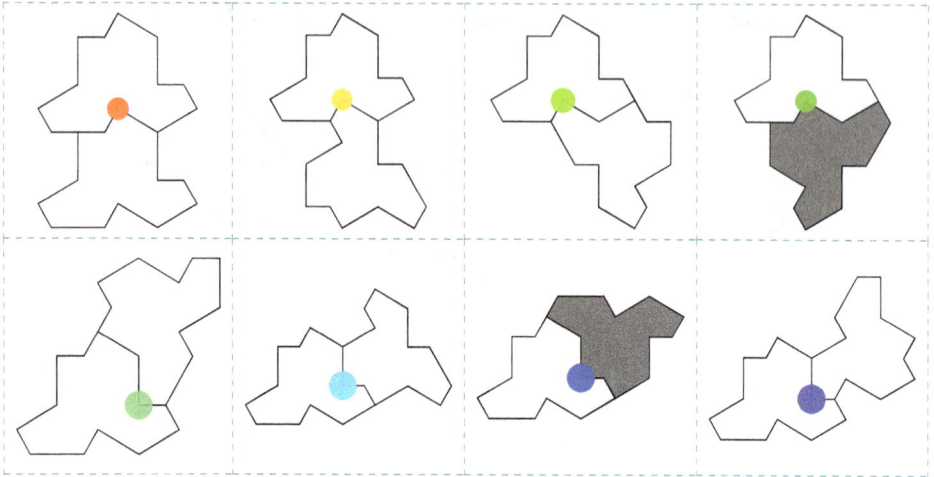

8 1/4+3/4 Vertex Configuration

Out[○]=

10 1/3+2/3 Vertex Configuration

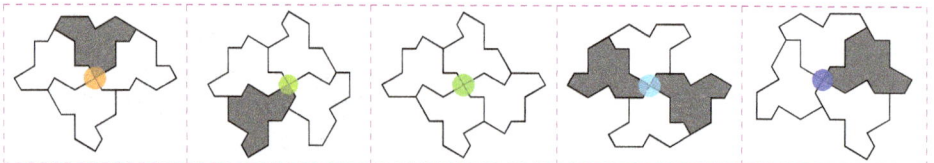

5 1/4 Vertex Configuration

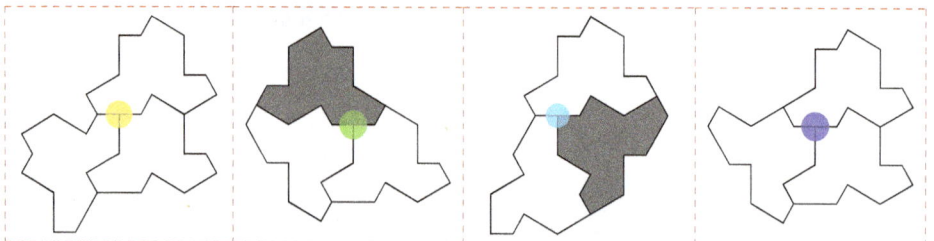

4 1/4+Edge "T" Configuration

Grow Cluster

Implementation of Grow Function (Code)

Visit wolfr.am/WSS2023-Ping.

Multi-growth Tree

From H2 to H7/H8

Take a two-hat tile, which is denoted by H2. Then we can find a way to grow it to H7, which is a meta-tile constructed from seven hat tiles. If we put all possible hats along the boundary point of H2, a multiway growth tree will be obtained:

```
In[ ]:= bp = boundaryPoints[H2];
        Labeled[
                Show[
                   Graphics[{PointSize[.03], Point[bp]}],
                   show[H2, EdgeThick → False]
                   ],
                Style[Column[H2], Purple, Bold],
            Right
        ]
```

Out[ ]=

$\{\{0, 0\}, 8, \text{False}, 2\}$
$\{\{0, 0\}, 2, \text{True}, 10\}$

```
In[ ]:= AbsoluteTiming[g = growthTree[H2, 3,
            SelectBoundaryPoints → bp, VisualizeNode → "Polygon"] // QuietEcho]
```

The result would be huge and would take 20 minutes to evaluate. Following is the iconized result to avoid this tedious evaluation:

```
In[ ]:= g = Graph[...] ;
```

Find all possible paths from H2 to H7:

```
In[ ]:= candidatesV = Select[VertexList[g], SubsetQ[#[[1]], canonicalHat @@@ H7] &];
        init = VertexList[g][[1]];
```

```
In[ ]:=  Graph[
            Union @@ (DirectedEdge[#1, #2] & @@@
                        Partition[First[FindPath[g, init, candidatesV[[#]]]], 2, 1] & /@ Range[10]),
            VertexShapeFunction →
              (Inset[show[#2[[1]], EdgeThick → False], #1, Center, 2 * #3] &),
            VertexSize → .5,
            ImageSize → Full,
            GraphLayout → "LayeredEmbedding"
          ]
```

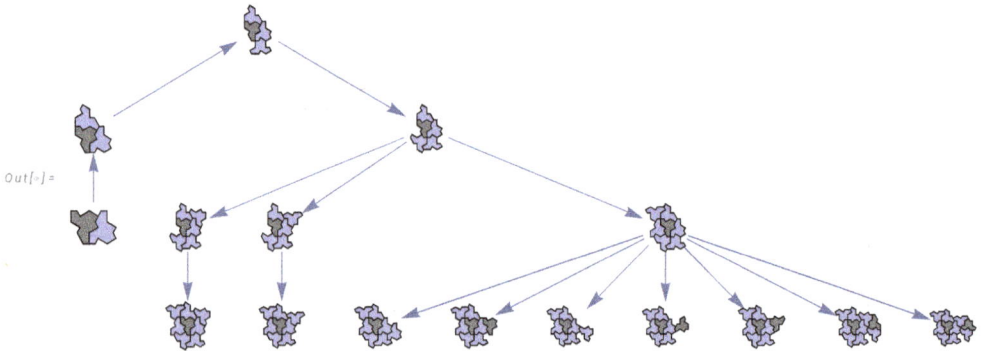

The two leaves on the lower-left side include H7, and the lower subtrees on the right-hand side all include H7. Then H7 is forced to H8:

```
In[ ]:=  With[{cp = getVertexPos[6, H7[[4]]]},
            g = growthTree[H7, 1, SelectBoundaryPoints → {cp}, VisualizeNode → True,
                EdgeShapeFunction →
                  ({Lighter @ Purple, Arrowheads[0.07], Arrow[#1, 0.3]} &),
                VertexShapeFunction →
                  (Inset[Show[{show[#2[[1]], EdgeThick → False], Graphics[
                            {Hue[.45], PointSize[.08], Point[cp]}]}], #1, Center, 2 * #3] &)
              ] // QuietEcho;
            Graph[g, ImageSize → Large]
          ]
```

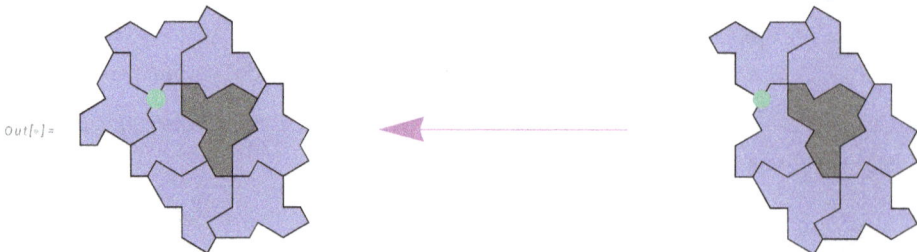

From Symmetric 11, 11, 11

The second-to-last vertex configuration of 1/3 + 1/3 + 1/3 is 11, 11, 11. This kind of configuration does not appear in any cluster, so here I try to prove it is invalid.

If a tile cluster is invalid, then its multiway growth tree is finite, which means it cannot grow to cover the whole plane without any holes.

The first step is to fill the three concave vertices, choosing points at the deepest position of each concave vertex:

```
In[ ]:= init = Prepend[#, {0, 0}] & /@ {{0, False, 11}, {4, False, 11}, {8, False, 11}};
```

$$pts = \{\{-\frac{1}{2}, \frac{\sqrt{3}}{2}\}, \{-\frac{1}{2}, -\frac{\sqrt{3}}{2}\}, \{1, 0\}\};$$

```
Labeled[
  Show[
    {show[init, ImageSize → Small], Graphics[{PointSize[.06], Yellow, Point[pts]}]}],
    Style[Column[init[[All, 2 ;; ]]], Blue, Italic, Bold],
    Right
  ]
```

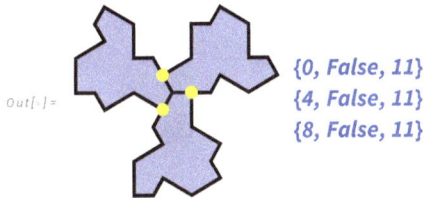

Out[ ]=
{0, False, 11}
{4, False, 11}
{8, False, 11}

```
In[ ]:= g = QuietEcho @
         growthTree[init, 3, SelectBoundaryPoints → pts, VisualizeNode → "Cluster"];
Graph[
  VertexReplace[g, Rule[#, #[[1]]] & /@ VertexList[g]],
  EdgeShapeFunction → ({Arrowheads[0.07], Arrow[#1, 0.3]} &),
  VertexShapeFunction → (Inset[show[#2], #1, Center, 2 * #3] &)
]
```

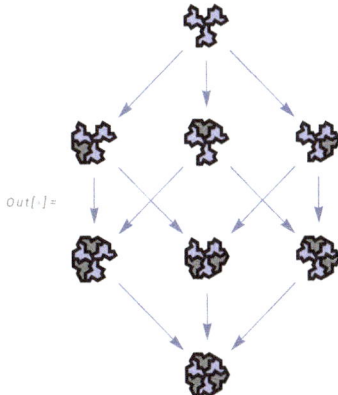

Out[ ]=

If the tree is finite, it will take a considerably long time to reach a fixed state. Otherwise, if it is infinite, it will never converge on a fixed state.

An example multiway tree is shown in the following:

```
In[ ]:= g = Graph[...] ;

Graph[g,
    GraphLayout → "LayeredEmbedding",
    VertexShapeFunction →
        (Inset[show[#2, EdgeThick → False], #1, Center, 2 * #3] &),
    EdgeShapeFunction → ({Arrowheads[0.01], Arrow[#1, 0.15]} &),
    ImageSize → Full,
    VertexSize → .65]
```

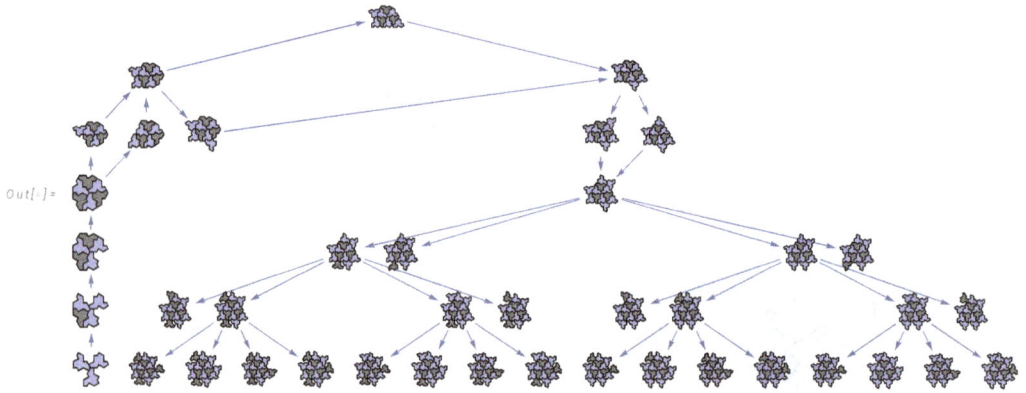

When the tree grows deeper, the number of concavities on the boundary gets larger. However, many of the gaps are not possible to fill. So this tree is finite, i.e. the 11, 11, 11 vertex configuration is invalid.

From Symmetric 6, 10, 6, 10

The fourth result of the 1/4 vertex conclusion is 6, 10, 6, 10, which doesn't appear in the big cluster and seems to be invalid. Here I try to grow it and prove invalidity:

```
In[ ]:= Labeled[
    show[Prepend[#, {0, 0}] & /@ oneFourthVertexTheoremLis[[4]], ImageSize → Small],
    Style[Column[oneFourthVertexTheoremLis[[4]]], Purple, Italic], Right]
```

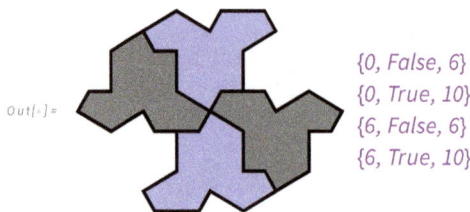

{0, False, 6}
{0, True, 10}
{6, False, 6}
{6, True, 10}

Then we try to grow it:

```
In[ ]:= init = Prepend[#, {0, 0}] & /@ oneFourthVertexTheoremLis[[4]];
       g = growthTree[init, 8, PointsNum → 1, VisualizeNode → "Cluster"] // QuietEcho;
```

```
In[ ]:= SimpleGraph[
         VertexReplace[g, Rule[#, #[[1]]] & /@ VertexList[g]],
         EdgeShapeFunction → ({Arrowheads[0.01], Arrow[#1, 0.2]} &),
         VertexShapeFunction →
           (Inset[show[#2, EdgeThick → False], #1, Center, 3 * #3] &),
         ImageSize → 1000
       ]
```

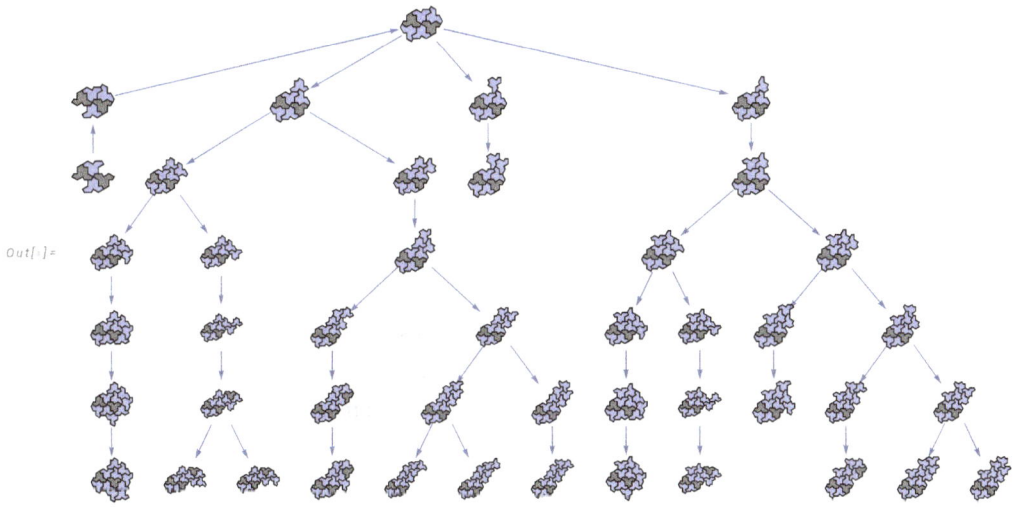

The leaves seem to grow into a diagonal line with two-hat width cutting the whole plane into two parts. There are no obvious invalid cases appearing here.

Here we try to grow more points on the boundary simultaneously:

```
In[ ]:= g = growthTree[init, 5, PointsNum → 2, VisualizeNode → "Cluster"] // QuietEcho;
```

```
In[ ]:=  SimpleGraph[
           VertexReplace[g, Rule[#, #[[1]]] &/@ VertexList[g]],
           EdgeShapeFunction → ({Arrowheads[0.008], Arrow[#1, 0.6]} &),
           VertexShapeFunction →
             (Inset[show[#2, EdgeThick → False], #1, Center, 5.5 * #3] &),
           GraphLayout → "LayeredDigraphEmbedding",
           ImageSize → 1000,
           AspectRatio → 1/3
         ]
```

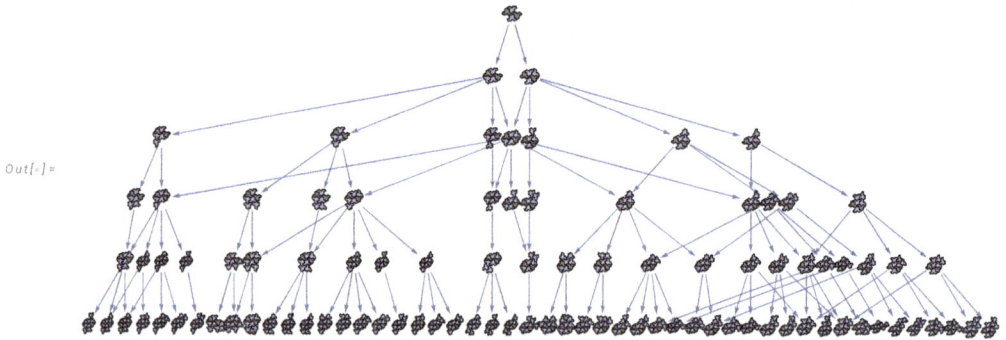

So far, there are no obvious dead ends, which means the multi-growth tree of 6, 10, 6, 10 is quite large, potentially infinite. If invalidity is to be proved, the tree will need to be larger and deeper. However, with branches growing more and more, the computation would be much too time-consuming to be practical.

Find Subclusters

Code for Finding Subclusters

Visit wolfr.am/WSS2023-Ping.

Examples

H7 and H8 are important meta-tiles that are the base of the substitution rule. There is a good way to find out all subclusters in a big cluster, or super-tile.

Here is a massive cluster, which is made up of 169 hat tiles:

```
In[ ]:=  cluster = {...} + ;
         cluster // Length
```

```
Out[ ]=  169
```

```
In[•]:=  Row[{
             Labeled[show[H7, ImageSize → Tiny],
                Style["H7", Purple, Bold, Italic], ContentSize → {120, 120}],
             Labeled[show[H8, ImageSize → Tiny],
                Style["H8", Purple, Bold, Italic], ContentSize → {140, 120}],
             Labeled[show[cluster, EdgeThick → False, ImageSize → Small],
                Style["cluster", Purple, Bold, Italic], ContentSize → {250, 200}]
         }]
```

Out[•]=

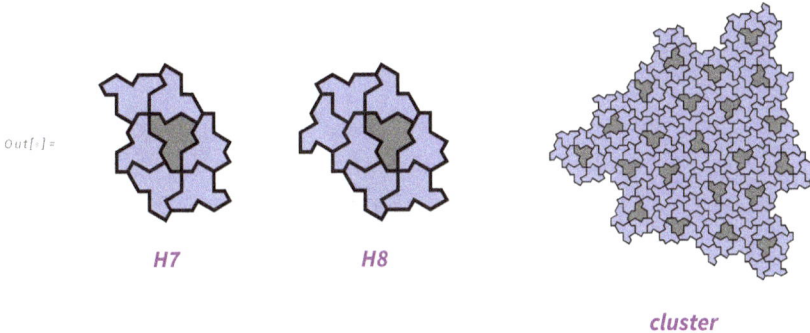

H7   *H8*

cluster

The idea is to give each edge a tag from 1 to 13 and record the special order of H7 and H8
boundary tags and their edge vectors (which are easily obtained from their topologic graphs):

```
In[•]:=  g = graphFromTiling[H7];

         Grid[
          Transpose[{
               {
                 Labeled[Graph[#1], Style["Topologic Graph for H7", Darker@Red, Italic],
                     Bottom], Labeled[Graph[#2],
                     Style["Topologic Graph for H8", Darker@Blue, Italic], Bottom]
               },
               {
                 Labeled[Graph[#1, VertexCoordinates → VertexList[#1]],
                     Style["Metric Graph for H7", Darker@Red, Italic], Bottom],
                 Labeled[Graph[#2, VertexCoordinates → VertexList[#2]],
                     Style["Metric Graph for H8", Darker@Blue, Italic], Bottom]
               },
               {
                 Labeled[
                  Show[
                   Graphics[
                    Replace[
                     H7,
                     {x_?(FreeQ[True]) :→ {White, EdgeForm[White], hat@@x},
                       x_?(!FreeQ[True][#] &) :→
```

```
                            {Lighter@Black, EdgeForm[White], hat@@x}}
                  , {1}]
              ],
              Graph[#1,
                VertexCoordinates → VertexList[#1], EdgeStyle → Directive[Red]]
            ],
            Style["H7 Underlay Graph", Darker@Red, Italic], Bottom],
          Labeled[
            Show[
              Graphics[
                Replace[
                  H8,
                  {x_?(FreeQ[True]) :→ {White, EdgeForm[White], hat@@x},
                    x_?(!FreeQ[True][#] &) :→
                      {Lighter@Black, EdgeForm[White], hat@@x}}
                  , {1}]
              ],
              Graph[#2,
                VertexCoordinates → VertexList[#2], EdgeStyle → Directive[Blue]]
            ],
            Style["H8 Underlay Graph", Darker@Blue, Italic], Bottom]
          }
        } &@@(graphFromTiling/@{H7, H8})
    ],
    Frame → All, FrameStyle → LightGray
  ]
```

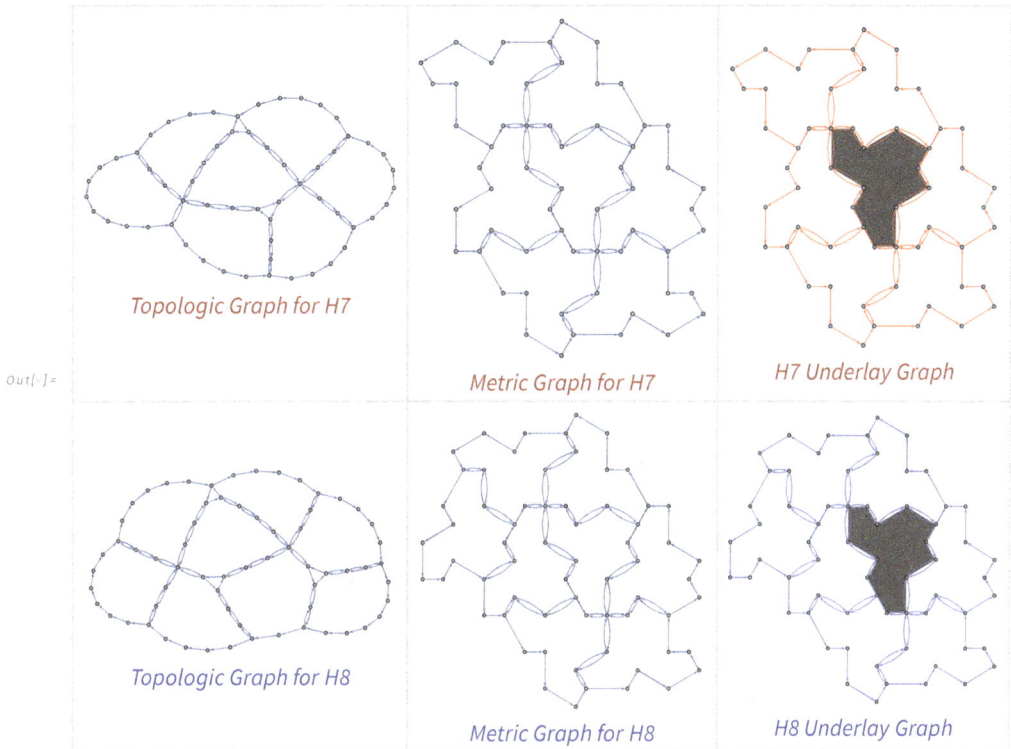

Topologic Graph for H7

Out[·]=

Metric Graph for H7

H7 Underlay Graph

Topologic Graph for H8

Metric Graph for H8

H8 Underlay Graph

There is an easy way to find subclusters with the built-in function FindCycle, as long as we know the edge tags' sequence. These are unique enough for each subcluster.

Notice that every H7 or H8 has a reflected hat in its center. Imagine walking from the sixth vertex of the reflected hat—the top point of the black tile—along the boundary, and it is easy to get the tag sequence of boundary edges.

Given the length, it is easy to find the cycle:

```
In[·]:= cycles = Association[];
       cycles["H7"] = First[FindCycle[graphFromTiling[H7], {37}, All]];
       cycles["H8"] = First[FindCycle[graphFromTiling[H8], {37}, All]];

       Row[
        MapThread[
         Labeled[Graph[#1, VertexCoordinates → #1[[All, 1]],
            ImageSize → 200, EdgeStyle → #3], #2] &,
         {
          {cycles["H7"], cycles["H8"]},
          {Style["H7 Boundary", Darker@Red], Style["H8 Boundary", Darker@Blue]},
          {Red, Blue}
         }], Invisible[ConstantArray[" ", 6]]]
```

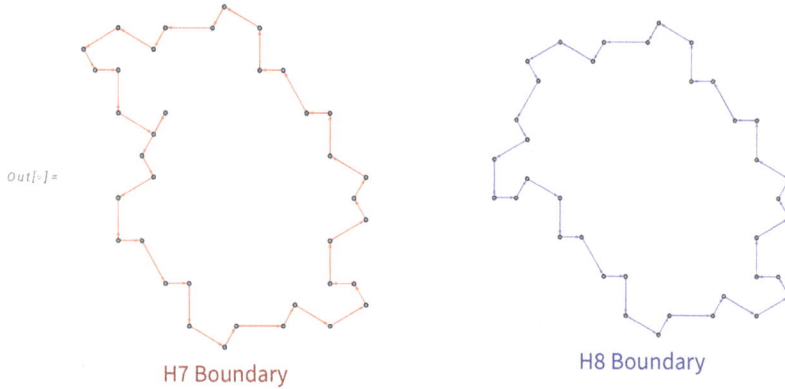

H7 Boundary

H8 Boundary

Additionally, each vector of the boundary edge can be recorded to nail down the problem and find all cycles faster:

```
In[•]:= AbsoluteTiming[cycleH7 = findByWalk["H7"][cluster];]
       AbsoluteTiming[cycleH8 = findByWalk["H8"][cluster];]
```

```
Out[•]= {0.099434, Null}
```

```
Out[•]= {0.0962224, Null}
```

```
In[•]:= Row[{
         Labeled[
           Show[
             show[cluster, EdgeThick → False, ChooseColor → {White, Lighter@Black}],
             Graph[#, VertexCoordinates → #[[All, 1]],
                   EdgeStyle → Directive[Thick, Red]] & /@ cycleH7,
             ImageSize → 250],
           Style["H7 for cluster", Red, Italic, Bold],
           ContentSize → {250, 250}],
         Labeled[
           Show[
             show[cluster, EdgeThick → False, ChooseColor → {White, Lighter@Black}],
             Graph[#, VertexCoordinates → #[[All, 1]],
                   EdgeStyle → Directive[Thick, Blue]] & /@ cycleH8,
             ImageSize → 250],
           Style["H8 for cluster", Blue, Italic, Bold],
           ContentSize → {250, 250}
           ]
         }
        ]
```

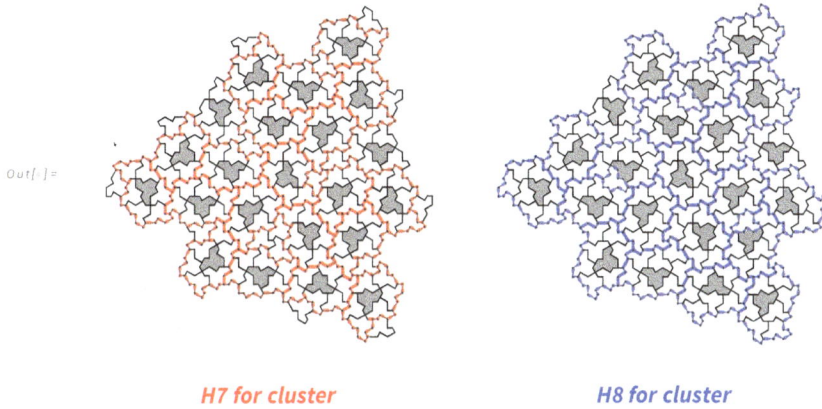

Out[ ]=

H7 for cluster   *H8 for cluster*

From Four Hat Tiles to Super-Tile

It is surprising that, with the grow function and growth tree, a path from the second and last cases of vertex configuration 1/4 + 1/4 + 1/4 + 1/4 to the H7/H8 super-tile is found, and is generated with vertex configurations rather than a substitution rule.

The dynamic growth is shown as follows:

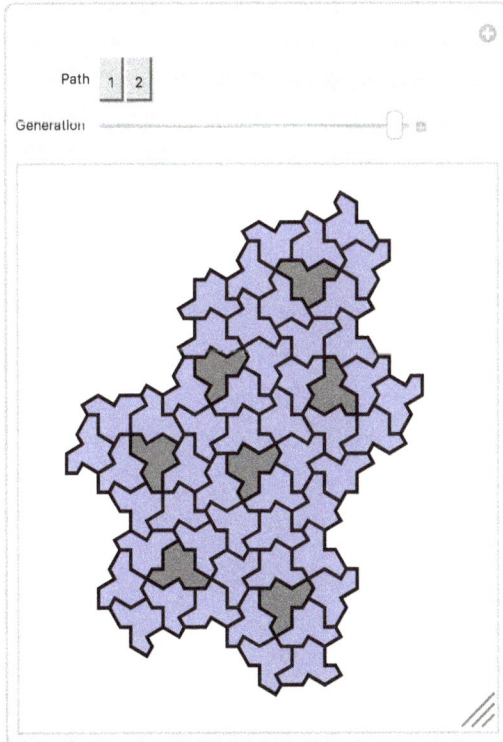

As for the other three cases, they do not appear in any big clusters so far, which is potentially eventually invalid. To prove their invalidity is a part of the work in the future.

Concluding Remarks

In summary, a fundamental growth function for the hat tile is obtained, which is fast enough to generate all possible valid and promising cases for a given cluster at a given certain point. Also, a multiway function is implemented to further study growth branches of a certain hat-tile cluster.

However, the algorithm is only built on the information of one point, which is extremely local and limited. If growth is from all points on the boundary, the time consumption is huge and the multiway tree is too complex to study.

The next step is to study the relationship between different vertex configurations, especially those that show up as neighbors in the big cluster. More information will generate fewer branches and make the multiway tree smaller and clearer.

Acknowledgments

I want to thank Brad Klee for his careful guidance and mentoring, and who also provided us with the important function PlanarPolygonFragmentation. It is really enjoyable to work with Johannes Martin. Thanks to Mark Greenberg for his help with graph layout codes and to Bob Nachbar for his help with pattern-matching codes. I'm glad to have met Zsombor Zoltán Méder, Ben Peter, Russell Martinez and everybody in the Wolfram Summer School. It was an unforgettable experience for me. Thanks!

References

1. D. Smith, et al. (2023), "An Aperiodic Monotile," arXiv preprint. arxiv.org/abs/2303.10798.

2. B. Klee (2023), "Hat Tilings via HTPF Equivalence," *Wolfram Community*. community.wolfram.com/groups/-/m/t/2858759.

3. B. Klee (2023), "Hat Combinatorics: The Ten Vertex Theorem," *Wolfram Community*. community.wolfram.com/groups/-/m/t/2935078.

Cite This Notebook

"Hat Tiling Space Reduction and Grow Function Implementation"
by Bowen Ping
Wolfram Community, STAFF PICKS, July 11, 2023
community.wolfram.com/groups/-/m/t/2957268

LLM-Powered Reviews of Student Work Samples

AARON CARVER

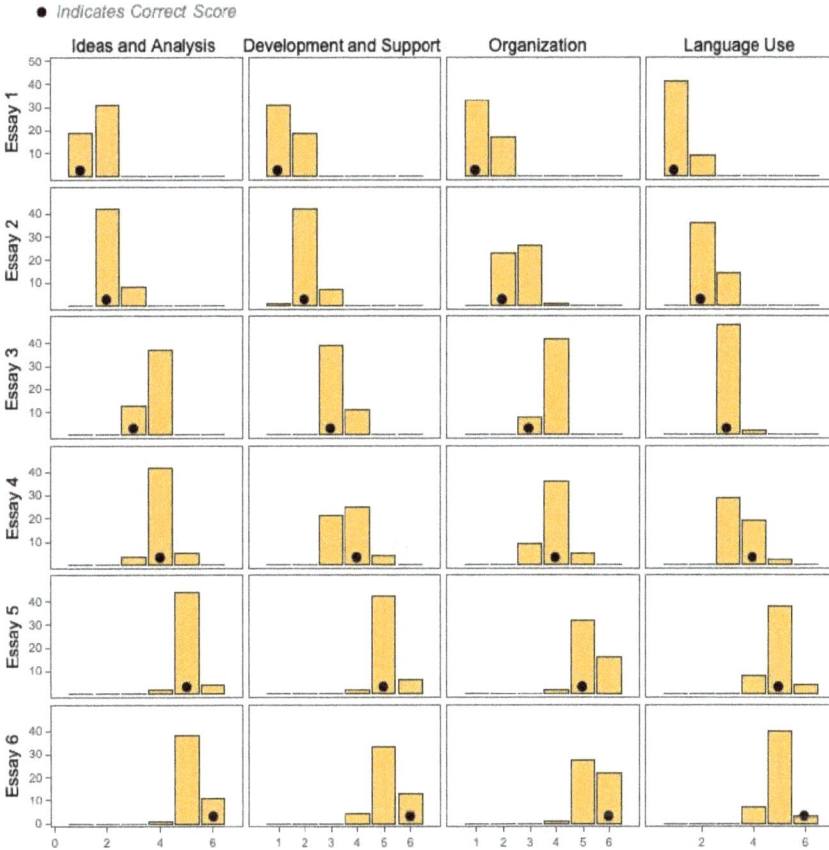

This project uses LLMs to review student work samples, specifically ACT writing essays. First, I explore using LLMs to apply a descriptive grading rubric that maps descriptions of writing quality to quantitative scores. Second, I use LLMs and the approach of few-shot learning to perform grammatical error correction (GEC), which is the process of identifying and fixing errors in the categories of spelling, punctuation, grammar and word choice.

For most of the writing samples, the LLM was able to accurately assess the correct score; however, each test produced some inaccurate and inconsistent results. I see several opportunities for further experimentation, which may lead to performance adequate for practical use such as varying temperature, preening effective examples for few-shot learning and implementing multiple calls to LLMs to disambiguate scores that vary by a single point.

For the GEC task utilizing few-shot learning, the LLMs showed a generally powerful ability to find errors and return them in a structured data format. However, the LLMs also "hallucinated" several false errors. In practice, a combination of traditional GEC tools with LLM-powered reviews will likely lead to the best results.

LLMs currently have the power and flexibility to provide useful reviews of student work samples, but the tools that I believe will succeed in practice will be those that add additional computational power and human judgement to LLM-powered workflows.

Applying a Descriptive Evaluation Rubric

For the ACT writing test, students are instructed to write an essay in response to a prompt. Two trained (human) evaluators score the essays from 1 to 6, in increments of 1, across each of four categories: (1) Ideas and Analysis; (2) Development and Support; (3) Organization; and (4) Language Use and Conventions. The overall score is the rounded sum of the four category scores.

- Note that for reporting purposes, the ACT score is reported as the sum of the essay score as graded by two separate graders, thus the reported range includes integers from 2 to 12. If the two graders report scores differing by more than 1, a third grader is consulted to resolve the scoring discrepancy (although the exact procedure for resolution is not reported on the ACT website).

In this experiment, I use six ACT writing sample essays and the official evaluation rubric, all provided by the creator of the test, ACT, Inc. Each of the samples is numbered from 1 to 6. The sample numbers used in the following **correspond with the essay's actual score**, ranging from **1 (lowest) to 6 (highest)**.

To access the essays or rubric, follow these links:
wolfr.am/Carver-Essays
wolfr.am/Carver-Rubric

Assigning an Overall Score

Each essay is sent in a separate request to the LLM, which is instructed to return an overall score from 1 to 6, but no rubric is provided. Without the rubric, the LLM doesn't distinguish at all by actual score and seems to "just pick" 4, regardless of the quality of the essay sample. While I found this surprising, I do appreciate that the LLM had such consistent behavior as a starting point, allowing for a clear signal of improvement in future experiments.

The real reason for choosing 4 is unclear, but perhaps the LLM "learned" this from a common human behavior when rating quality on integer scales. For example, the popular net promoter score (NPS) metric uses results from a survey in which users of a product rate the product on an integer scale from 1 to 10. Scores of 7 and 8 are so common that users of the NPS metric ignore those scores, as they represent "passives" who do not have a strong opinion. Perhaps the LLM "decided" to take a similarly passive approach.

Essay scores, no rubric (*N* = 50, GPT-3.5 Turbo, temperature = 1):

In[ ]:= **LLMFunction[**

     "Report a single score for this essay from 1 to 6 in this format: score", Expression]

● *Indicates Correct Score*

Out[∘]=

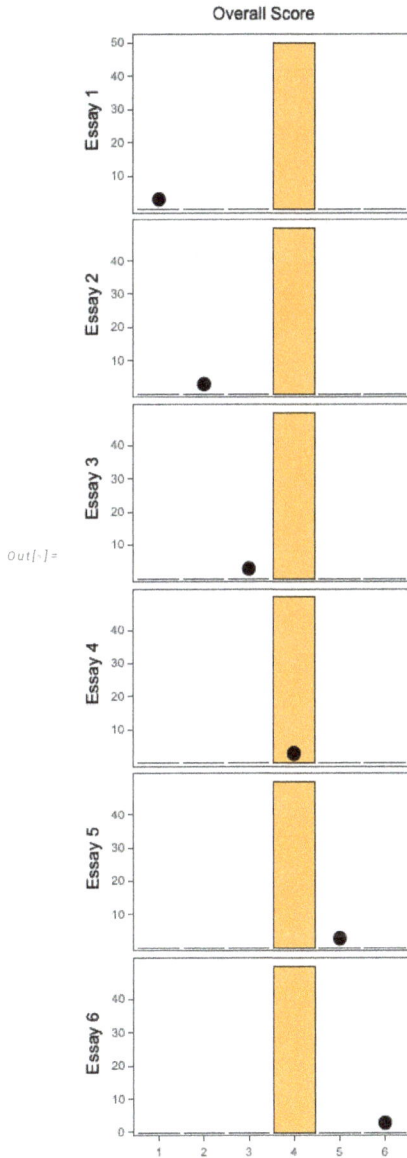

The same experiment is run again, but this time, the rubric is provided to the LLM at inference time. The results show that the LLM was able to learn how to grade the essays, but the results are somewhat inaccurate and inconsistent.

Essay scores, full rubric (*N* = 50, GPT-3.5 Turbo, temperature = 1):

LLMFunction[

    "You are the best professional evaluator for ACT writing samples. You are consistent,
        fair, and thorough in your evaluations. Use the following rubrics as guides.
        Report a single score for the whole essay from 1 to 6 in this format: score.
        Do not report a score for each category – this is very important – don't
        report a per–category score, just report a score for the entire essay."
      <> "Ideas and Analysis: " <> ToString[ideasAndAnalysisRubric]
      <> "Development and Support: " <> ToString[developmentAndSupportRubric]
      <> "Organization" <> ToString[organizationRubric]
      <> "Language Use" <> ToString[languageUseRubric]
      <> "Essay:``", Expression,
    LLMEvaluator → <| "Temperature" → 1, "Model" → "gpt–3.5–turbo" |>]

● *Indicates Correct Score*

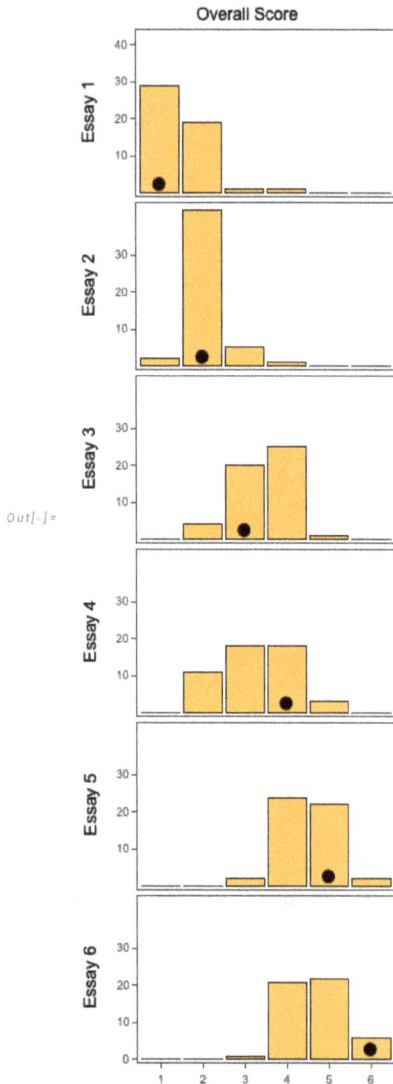

Assigning Category Scores

In an attempt to increase accuracy and consistency, the LLM is asked to score the essays by category. The LLM was quite effective in evaluating the low-scoring samples (Essay 1, Essay 2), but struggled with the high-scoring samples (Essay 5, Essay 6).

Per-category scores, full rubric (N = 50, GPT-3.5 Turbo, temperature = 1):

```
In[*]:=  LLMFunction["You are the best professional evaluator for ACT writing samples. You are
                consistent, fair, and thorough in your evaluations. Report a single
                score for each category, from 1 to 6, for the essay by applying the
                following rubrics. Report the scores in this format: <|\"Ideas and
                Analysis:\" -> \"score\", \"Development and Support\" -> \"score\",
                \"Organization\" -> \"score\", \"Language Use\" -> \"score\"|>"
            <> "Ideas and Analysis: " <> ToString[ideasAndAnalysisRubric]
            <> "Development and Support: " <> ToString[developmentAndSupportRubric]
            <> "Organization" <> ToString[organizationRubric]
            <> "Language Use" <> ToString[languageUseRubric]
            <> "Essay:``", Expression,
        LLMEvaluator -> <| "Temperature" -> 1, "Model" -> "gpt-3.5-turbo" |>];
```

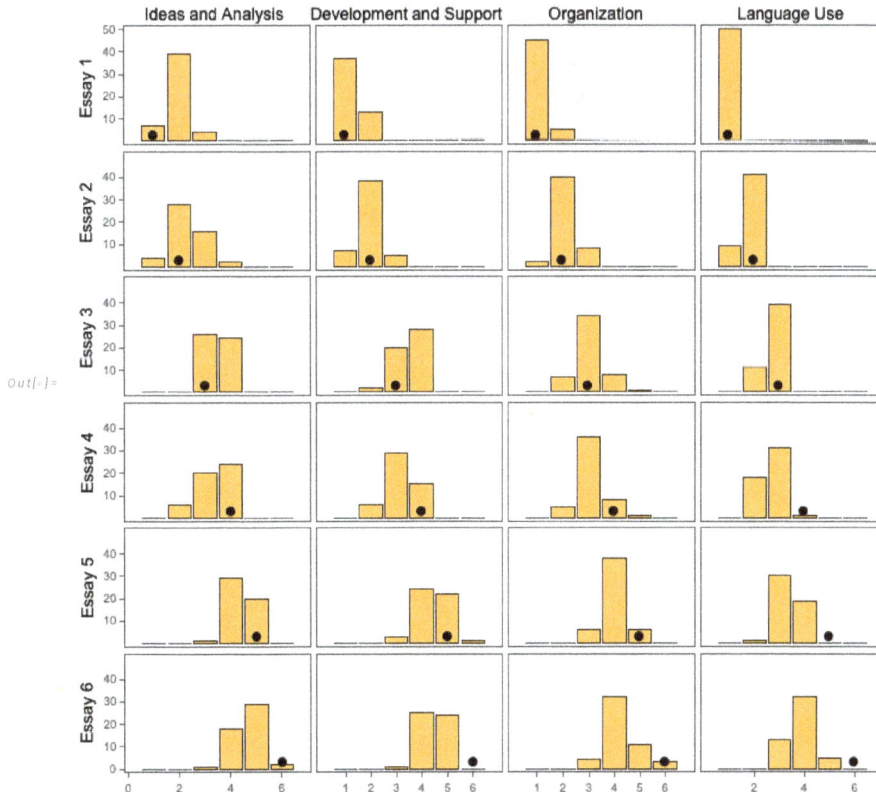

The experiment was run again with GPT-4, which resulted in better performance.

Per-category scores, full rubric (*N* = 50, GPT-4, temperature = 1):

In[·]:= **LLMFunction["You are the best professional evaluator for ACT writing samples. You are consistent, fair, and thorough in your evaluations. Report a single score for each category, from 1 to 6, for the essay by applying the following rubrics. Report the scores in this format: <|\"Ideas and Analysis:\" –> \"score\", \"Development and Support\" –> \"score\", \"Organization\" –> \"score\", \"Language Use\" –> \"score\"|>"**
    <> "Ideas and Analysis: " <> ToString[ideasAndAnalysisRubric]
    <> "Development and Support: " <> ToString[developmentAndSupportRubric]
    <> "Organization" <> ToString[organizationRubric]
    <> "Language Use" <> ToString[languageUseRubric]
    <> "Essay:``", Expression,
   LLMEvaluator → <| "Temperature" → 1, "Model" → "gpt–4" |>];

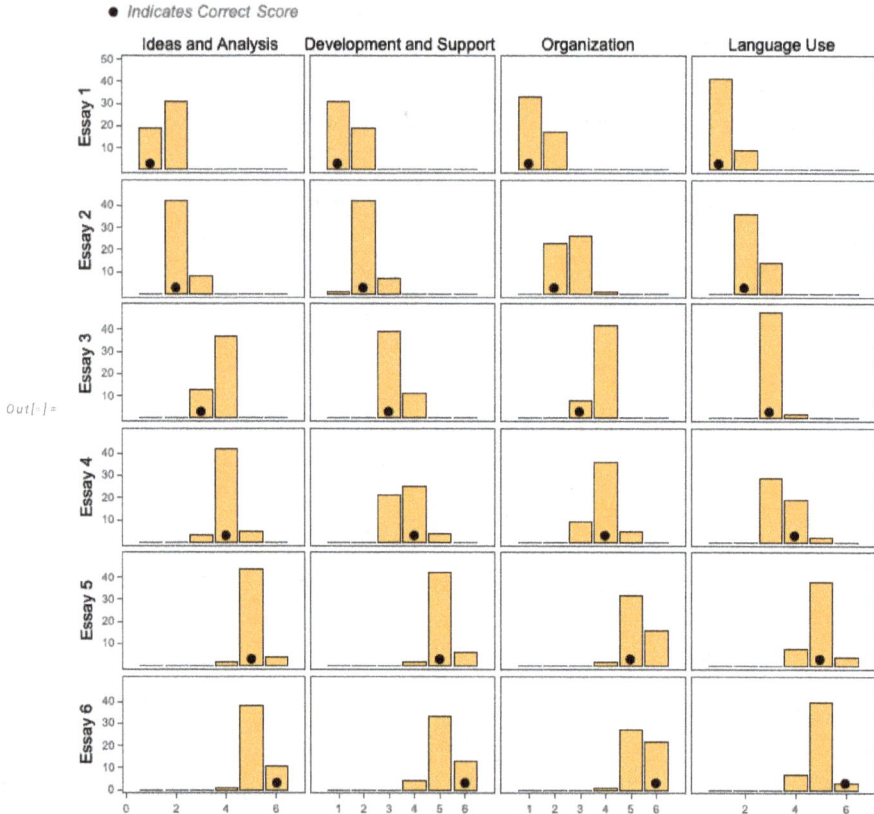

Again using GPT-4, the LLM's "temperature" was lowered to 0. Results were more consistent, but quite inaccurate for Essay 6.

Per-category scores, full rubric (*N* = 25, GPT-4, temperature = 0):

In[ ]:= **LLMFunction["You are the best professional evaluator for ACT writing samples. You are**
consistent, fair, and thorough in your evaluations. Report a single
score for each category, from 1 to 6, for the essay by applying the
following rubrics. Report the scores in this format: <|\"Ideas and
Analysis:\" –> \"score\", \"Development and Support\" –> \"score\",
\"Organization\" –> \"score\", \"Language Use\" –> \"score\"|>"
<> "Ideas and Analysis: " <> ToString[ideasAndAnalysisRubric]
<> "Development and Support: " <> ToString[developmentAndSupportRubric]
<> "Organization" <> ToString[organizationRubric]
<> "Language Use" <> ToString[languageUseRubric]
<> "Essay:` `", Expression ,
LLMEvaluator → <| "Temperature" → 0, "Model" → "gpt–4" |>];

Out[ ]=

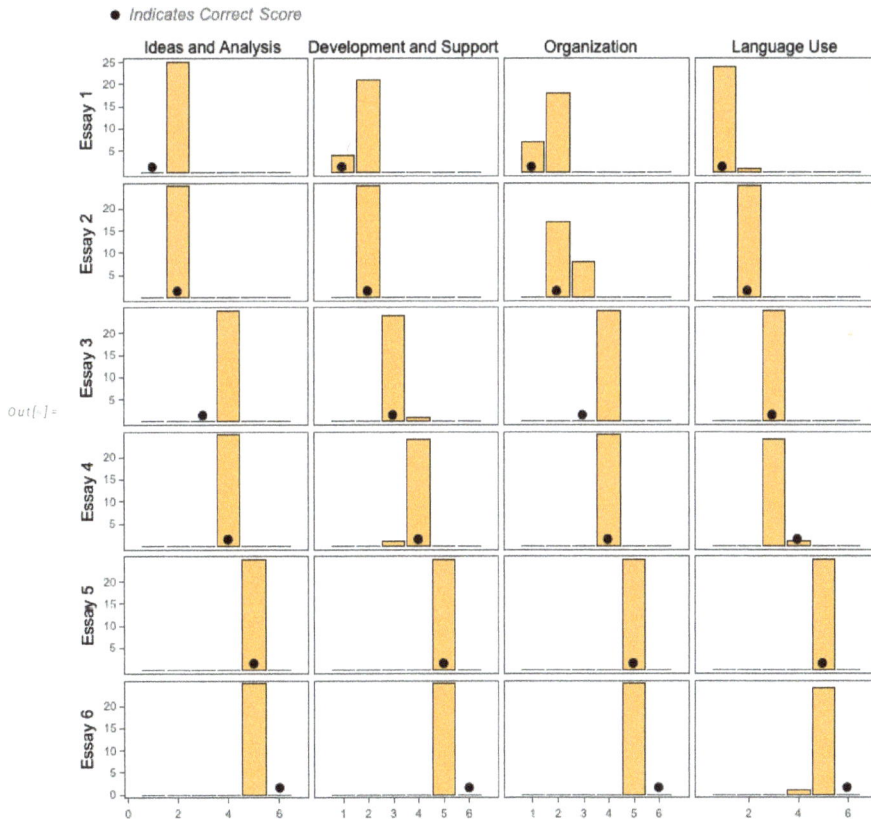

Just by using GPT-4 with a temperature of 0, we achieve results that I suspect rival those of trained groups of human graders—with the exception of assigning scores to Essay 6. How might one fix this in practice? Perhaps the essay can be sent to a second LLM that specifically has instructions for how to disambiguate essays of scores 5 and 6. Although the regular human workflow for evaluating these essays includes a single review by two separate humans, the use of LLMs allows for significantly more sophisticated workflows that can involve an arbitrary number of LLM requests.

After running these experiments, I'm struck by two conclusions: (1) LLMs have a powerful ability to follow inference-time instructions for scoring student work samples; and (2) to get results that are good enough to use in practice, one needs to carefully test and curate prompts and workflows that can achieve greater-than-human performance.

Other Types of Evaluations

I chose to experiment with the task of using a descriptive rubric to evaluate student essays because it's a particularly challenging task. There's not a "correct" answer, just essays that may or may not have responses with certain qualities like strong "Ideas and Analysis" and proper "Language Use." Even with this somewhat ambiguous task and inference-time instructions, the LLM was able to produce human-like results.

For evaluating other types of student work samples, such as short-answer responses that correspond to questions that do have correct answers, I suspect that the LLM will rather easily provide human-like or better-than-human results. For example, the teacher may have students write short-answer (1–2 sentence) responses to a set of 20 questions about a particular topic. While grading the responses, the teacher will have in their head some context about what concepts represent a correct answer and what misconceptions are most common in incorrect answers. The teacher will also have a set of heuristics for how many numerical points each concept is worth, how many points by which to reduce a score for small errors, etc.

If a teacher were to write out this context, a prompt could be created that knows about what the correct answer is, what concepts are involved, what mistakes are common and how to deduct points for different types of mistakes. Given a product that allows users to give this sort of input, one can imagine allowing the teacher to have the LLM grade each question by each student and return scores and commentary for each. If needed, the tool could then also show distributions of performance for different questions or concepts, different student populations, etc.

Grammatical Error Correction (GEC)

Traditional GEC systems, often referred to as "spellcheck" software, use a combination of replacement rules, such as for misspelled words, and machine learning models trained on datasets of text samples with errors. Typically, these text samples are created by starting with a grammatically correct sentence and introducing a mistake, recording that mistake as a "tag" for the machine learning training process, and testing the GEC system on a corpus of millions of such examples.

This type of spellcheck software is effective across many types of errors, but when several errors are seen in combination, such as in a low-quality writing sample, the spellcheck software struggles to identify compositions of errors and possible solutions. These GEC systems are also not able to use semantic meaning as context within a text sample, which can lead to both false negative and false positive mistakes.

Local "Error Context" and Returning Structured Data from an LLM

One difficulty of measuring the performance of GEC software for student work samples is that no existing system can be relied upon to find all errors. Therefore, I manually created a list of each error and a suggested fix for all six essays. A second challenge is that because the LLM is instructed to return the error within a Wolfram Language rule structure, the LLM must choose an "error context" that may differ from the one I chose. For example, the same error could be fixed using different rules where the "error context" is either *they're book* or just *they're*:

"they're book" → "their book"

Or, equivalently:

"they're" → "their"

In order to determine whether a GEC program found and corrected the error, I used two approaches for Essay 1. The first approach was to manually review the results returned by the LLM at $K = \{1, 5, 10, 20\}$. The second was to match the "error context" values from the manually created key to the values returned by the LLM. As seen in the plot for Essay 1, the automated detection generally performed quite well below $K = 20$.

This matching of "error context" demonstrated the ability of the LLM to follow the examples provided to it for few-shot learning. The LLM correctly identified the error context for simple situations, such as misspelled words by returning just the word, and for situations in which two or more words constituted the expected "error context." This automated detection was used for estimating the GEC performance for the remaining essay samples.

GEC Results vs. K (the Number of Few-Shot Learning Examples) across Essays 1 to 6

However, by using GPT-3.5 Turbo and increasing K (the number of examples provided at inference time for few-shot learning), results comparable to a modern spellcheck system were obtained.

False negatives and positives per real error vs. K:

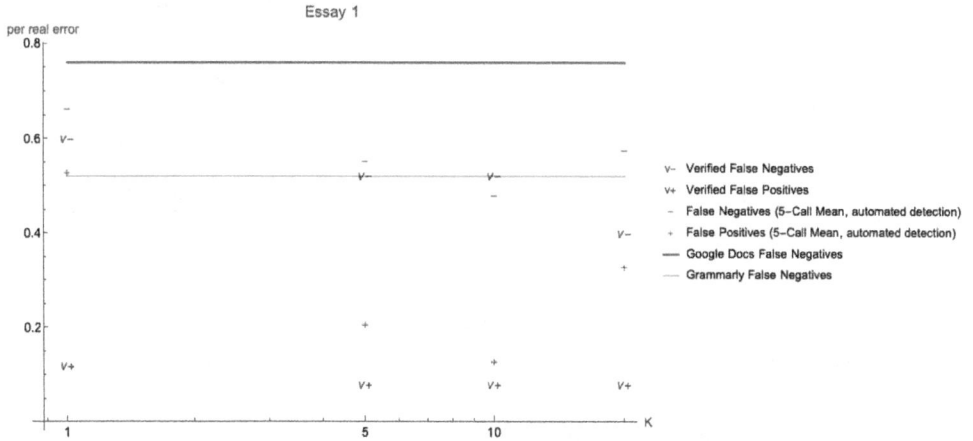

Essay 1

For Essays 2 through 4, the LLM showed similar or worse performance than popular commercial GEC software.

False negatives per real error vs. K:

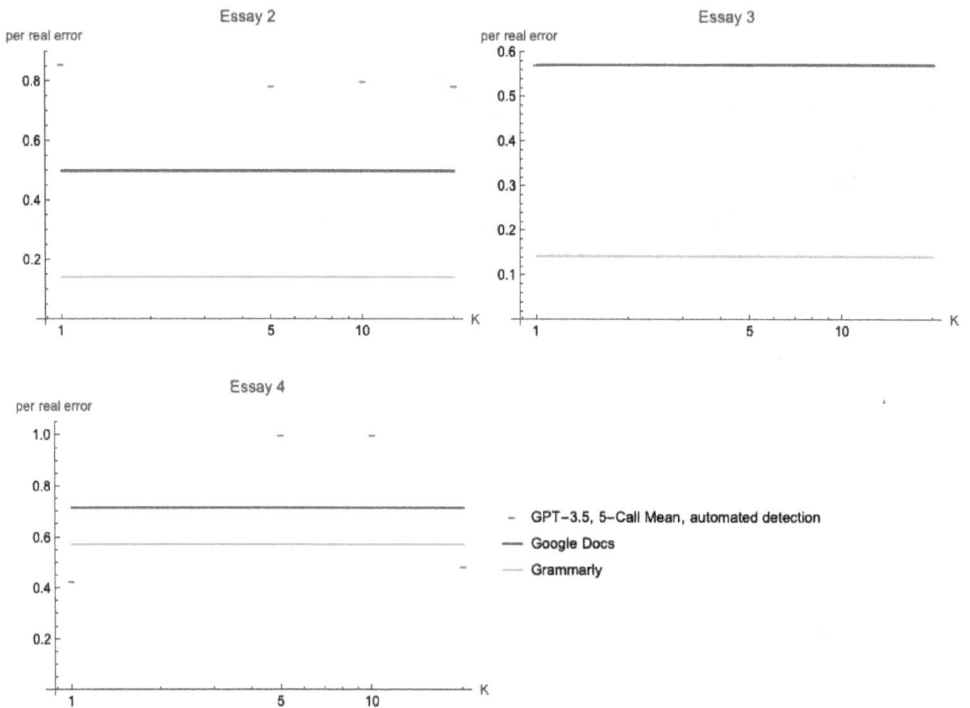

Essay 2

Essay 3

Essay 4

For the high-quality writing samples, the LLM was not able to find errors as well as commercial GEC software.

False negatives and positives per real error vs. K:

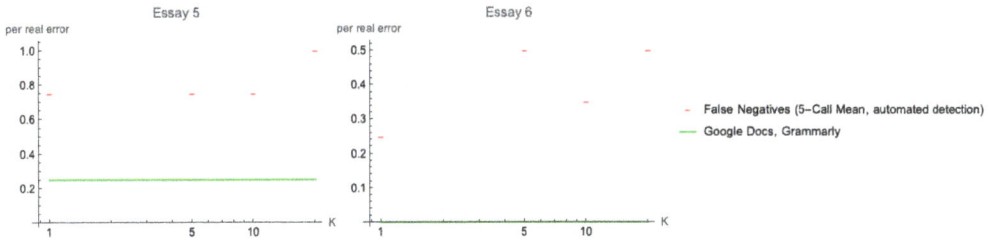

The LLM demonstrated the ability to return data in a structured format when provided with examples for few-shot learning. However, the lack of identification of common grammatical and spelling errors, especially in high-quality writing samples, suggests that the LLM might be best used as an aid to a rules-based approach to GEC.

Overall, these GEC experiments were very preliminary in terms of the use of LLMs. Further experimentation of varying the temperature, the prompt, the examples used for few-shot learning and the use of multiple LLM calls in various sequences may provide greatly improved performance. Additionally, due to the LLM's ability to successfully follow instructions for returning structured data with appropriate context, tools such as a "fact-checker" could be implemented in a similar manner as the GEC solution presented here.

UI Markup

For the task of showing a user what changes were made, an early mockup of a UI is shown here. The main method for displaying the changes made is the use of Wolfram Language's ability to perform sequence matching for string characters, as implemented in the resource function HighlightTextDifferences (from the Wolfram Function Repository).

This UI is illustrative, but could be significantly improved by allowing the user to accept or reject a change, by showing the sequences more clearly and by giving a summary of the types of errors found.

```
In[ ]:=   essay1 = ⌈ "..." ⊹ ⌉;

In[ ]:=   essay1Fixed = ⌈ "..." ⊹ ⌉;
```

Use HighlightTextDifferences to display text edits to the user:

In[·]:= Column[{Text[Style[essay1, "TextStyling"]],
      ResourceFunction["HighlightTextDifferences"][essay1, essay1Fixed],
      Text[Style[essay1Fixed, "TextStyling"]]}, Frame → All]

Well Machines are good but they take people jobs like if they don't know how to use it they get fired and they'll find someone else and it's more easyer with machines but sometimes they don't need people because of this machines do there own job and there be many people that lack on there job but the intelligent machines sometimes may not work or they'll brake easy and it's waste of money on this machines and there really expensive to buy but they help alot at the same time it help alot but at the same time this intelligent machines work and some don't work but many store buy them and end up broken or not working but many stores gets them and end up wasting money on this intelligent machines' but how does it help us and the comunity because some people get fired because they do not need him because of this machines many people are losing job's because of this machines.

Out[·]= Well, ~~M~~machines are good, but they take people's jobs ~~like.~~ ~~i~~If they don~~'~~t know how to use ~~i~~them, they get fired, a~~nd they'll f~~ind someone else ~~and~~will be hi~~t's~~red. ~~more~~It's easy~~i~~er with machines, but sometimes they don~~'~~t need people because ~~of~~ th~~i~~ese machines can do the~~re own~~ job a~~ond~~ the~~i~~re ~~be~~own. ~~m~~Many people th~~la~~tck ski~~lack~~lls fo~~nr~~ the~~i~~re job~~b~~uts, ~~the~~and intelligent machin~~es sometim~~es may not ~~always~~ work or ~~they~~may~~'ll~~ break~~e~~ eas~~y~~ and~~ily.~~ ~~i~~It~~'~~s a waste of money ~~on~~ th~~iso~~ ~~machin~~invest ~~ai~~nd the~~r~~se ~~really~~ expensive ~~to buy~~machines. ~~but~~However, they can be help~~a~~ful~~ot~~ at the same time. ~~it~~They help a lot, but at the same time, ~~this~~some intelligent machines work ~~and~~while ~~som~~others don~~'~~t ~~work but~~ ~~.~~ ~~m~~Many stores buy them, ~~a~~but they ~~oft~~end ~~end~~ up broken or not working ~~but many.~~ ~~s~~Stores ~~gets them and end up~~ wast~~ing~~e money on th~~i~~ese intelligent machines~~'~~ ~~but.~~ ~~h~~How does it help us and the community ~~because~~? ~~s~~Some people get fired because they ~~do~~are not ~~lo~~no ~~ge~~longe~~d~~r ~~him~~ ~~b~~neede~~ca~~d ~~du~~se t~~of~~ th~~i~~ese machines. ~~m~~Many people are losing job~~'~~s because of th~~is machines~~em.

Well, machines are good, but they take people's jobs. If they don't know how to use them, they get fired, and someone else will be hired. It's easier with machines, but sometimes they don't need people because these machines can do the job on their own. Many people lack skills for their jobs, and intelligent machines may not always work or may break easily. It's a waste of money to invest in these expensive machines. However, they can be helpful at the same time. They help a lot, but at the same time, some intelligent machines work while others don't. Many stores buy them, but they often end up broken or not working. Stores waste money on these intelligent machines. How does it help us and the community? Some people get fired because they are no longer needed due to these machines. Many people are losing jobs because of them.

Concluding Remarks

These experiments tested two different tasks: evaluation using a descriptive rubric and using GEC. In both, the LLM's performance showed remarkable flexibility to perform tasks as instructed at inference time, but lacked the level of accuracy and consistency needed for widespread use as a product. These results illustrate the need for computational power within product workflows both before and after the LLM calls in order to bridge the gap between the "out-of-the-box" LLM performance and users' needs.

Today, we typically have only "raw" interactions with the LLM as a processing "kernel," but I expect to soon be able to work with much more sophisticated and useful tools and workflows that have this LLM "kernel" as one very powerful and interesting processing step. I expect that there are many improvements to be found both by doing "LLM science" to study the behavior of these AI systems and by developing creative, simple-to-use products that allow users the ability to easily perform tasks that previously required intensive application of focused human thought.

Acknowledgments

Thanks to the many Wolfram Summer School staff members and students who helped with this project.

Sotiris Michos—for thinking through experimental approaches, reviewing many drafts of notebooks, providing encouragement and celebration and setting up discussions related to this project with varied groups.

Fez Zaman—for helping with data wrangling and visualization and discussing the potential use of multiway paths for grammatical error corrections.

Ghassane Aniba—for discussing experiments to run and prompts to use.

Bob Nachbar—for help generating graphics.

Eric Parfitt—for helping refactor code and creating sensible result plots.

Paul Abbott and Mark Greenberg—for discussion on the use of LLMs for teaching and many ideas for practical extensions of the areas explored in this project.

Jofre Espigulé Pons—for providing a particularly helpful guide to LLM functionality in Wolfram Language and discussing prompting techniques.

Christopher Wolfram—for discussing details of the LLMTool implementation and intuition about the interactions between LLMs and Wolfram Language.

References

1. A. Carver (2023), "LLM-Powered Reviews of Student Work Samples," *Wolfram Notebook Archive* [full set of experiments in notebook form]. notebookarchive.org/2023-07-5kv909m.

2. Writing Sample Essays [website]. www.act.org/content/act/en/products-and-services/the -act/test-preparation/writing-sample-essays.html.

3. The ACT Writing Test Scoring Rubric [website]. www.act.org/content/dam/act/unsecured /documents/Writing-Test-Scoring-Rubric.pdf.

Cite This Notebook

"LLM-Powered Reviews of Student Work Samples"

by Aaron Carver

Wolfram Community, STAFF PICKS, July 12, 2023

community.wolfram.com/groups/-/m/t/2958774

From Elliptic Curves to Diophantine Equations: A Journey through Rational Points

ADITI KULKARNI

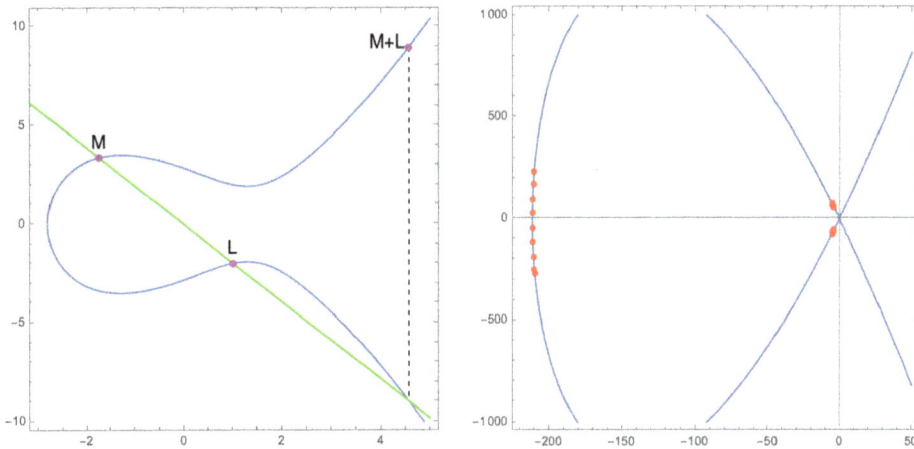

Elliptic curves, commonly represented as $y^2 = x^3 + A\,x + B$ (where A and B are constants) involve interactions among geometry, number theory and algebra. This project explores rational points on elliptic curves. By repeatedly applying a geometric procedure (the tangent-secant method) to a finite set of solutions, all of the possibly infinitely many rational solutions to an elliptic curve equation may be found. We will examine the relationship between integer solutions of Diophantine equations and rational points on elliptic curves. The simple-looking Diophantine equation $\frac{a}{b+c} + \frac{b}{a+c} + \frac{c}{a+b} = n$ (where n is a fixed positive integer) is infamous for having very large minimal solutions. This homogenous equation in three variables can be transformed to an elliptic curve in a two-dimensional plane. Additionally, rational points on the elliptic curve lead us to least positive integer solutions of the Diophantine equation. We will be learning about solutions of $\frac{a}{b+c} + \frac{b}{a+c} + \frac{c}{a+b} = 6$ in this project.

Background

Diophantine Equations

A Diophantine equation refers to an equation usually expressed as a polynomial with integer coefficients involving two or more unknowns. The primary focus of these equations lies in finding solutions that are limited to integers. There are several functions in Mathematica for solving equations, such as FindInstance, Solve, Reduce and FindRoot. A few examples include:

$In[•]:=$ **Solve[3 * x + 4 * y == 498, {x, y}, Integers]**

$Out[•]=$ $\left\{\left\{x \to \boxed{2 + 4\,c_1 \ \text{if} \ c_1 \in \mathbb{Z}}, y \to \boxed{123 - 3\,c_1 \ \text{if} \ c_1 \in \mathbb{Z}}\right\}\right\}$

$In[•]:=$ **FindInstance[x^2 + y^2 − 45 == z^2, {x, y, z}, Integers]**

$Out[•]=$ $\{\{x \to -8, y \to -9, z \to -10\}\}$

Diophantine equations can be of any degree. In Mathematica, there exists a specific function called FrobeniusSolve to solve first-degree/linear Diophantine equations (LDEs). Consider $12w + 39x + 25y + 7z = 207$; this LDE will be solved as:

$In[•]:=$ **FrobeniusSolve[{12, 39, 25, 7}, 207]**

$Out[•]=$ {{0, 0, 1, 26}, {0, 0, 8, 1}, {0, 1, 0, 24}, {1, 0, 5, 10}, {1, 1, 4, 8}, {1, 2, 3, 6},
{1, 3, 2, 4}, {1, 4, 1, 2}, {1, 5, 0, 0}, {2, 0, 2, 19}, {2, 1, 1, 17}, {2, 2, 0, 15},
{3, 0, 6, 3}, {3, 1, 5, 1}, {4, 0, 3, 12}, {4, 1, 2, 10}, {4, 2, 1, 8}, {4, 3, 0, 6},
{5, 0, 0, 21}, {6, 0, 4, 5}, {6, 1, 3, 3}, {6, 2, 2, 1}, {7, 0, 1, 14}, {7, 1, 0, 12}, {9, 0, 2, 7},
{9, 1, 1, 5}, {9, 2, 0, 3}, {11, 0, 3, 0}, {12, 0, 0, 9}, {14, 0, 1, 2}, {14, 1, 0, 0}}

Check one of the solutions:

$In[•]:=$ **ReplaceAll[12 w + 39 x + 25 y + 7 z, Thread[{w, x, y, z} → {4, 1, 2, 10}]] == 207**

$Out[•]=$ **True**

The nontrivial minimal solutions of a Diophantine equation are the smallest positive integers that satisfy the equation.

$\dfrac{a}{b+c} + \dfrac{b}{a+c} + \dfrac{c}{a+b} = n$ Where n Is a Fixed Positive Integer

This is the Diophantine equation that we will study in this project. Though it may look simple, this type of Diophantine equation is notorious for having a large minimal solution. It is computationally difficult to find a solution by brute force. Let's try for $n = 6$:

$In[\cdot]:=$ FindRoot[$\dfrac{a}{b+c} + \dfrac{b}{a+c} + \dfrac{c}{a+b}$ == n && y^2 == x^3 + A*x^2 + B*x, {x, y}]

$\cdots$ FindRoot : Value y in search specification {x, y } is not a number or array of numbers.

$Out[\cdot]:=$ FindRoot$\left[\dfrac{a}{b+c} + \dfrac{b}{a+c} + \dfrac{c}{a+b}$ == n && y^2 == $x^3 + Ax^2 + Bx$, {x, y}$\right]$

$In[\cdot]:=$ FindInstance[$\dfrac{a}{b+c} + \dfrac{b}{a+c} + \dfrac{c}{a+b}$ == 6, {a, b, c}, PositiveIntegers]

$\cdots$ FindInstance : The methods available to FindInstance are insufficient to find the requested instances or prove they do not exist.

$Out[\cdot]:=$ FindInstance$\left[\dfrac{c}{a+b} + \dfrac{b}{a+c} + \dfrac{a}{b+c}$ == 6, {a, b, c}, $\mathbb{Z}_{>0}\right]$

We will solve this using rational points on elliptic curves.

Elliptic Curves

An elliptic curve is defined over a field K and describes points in K^2, the Cartesian product of K with itself. The Weierstrass equation for an elliptic curve is defined as $y^2 = x^3 + Ax + B$ with some coefficients A and B in the field K:

$In[\cdot]:=$ Manipulate[ContourPlot[{y^2 == x^3 + a*x + b}, {x, −30, 70}, {y, −200, 200},
        Axes → True, PerformanceGoal → "Quality"], {a, −500, 500}, {b, −500, 500}]

$Out[\cdot]=$

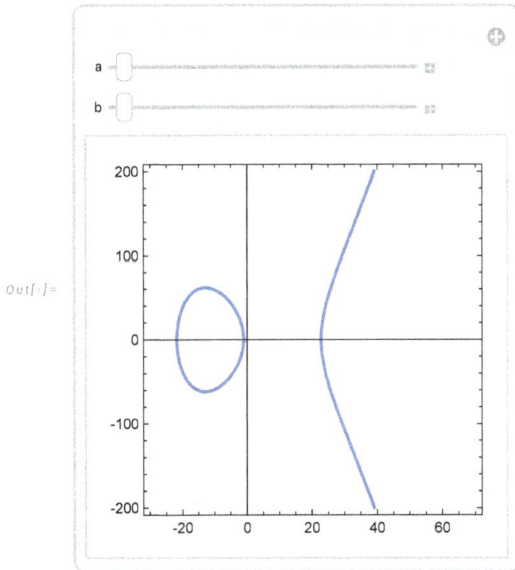

Here is the slightly modified and more general equation of an elliptic curve:

$$y^2 = x^3 + Ax^2 + Bx + C$$

Rational Points on an Elliptic Curve

A rational number is the quotient of two integers. A point (x, y) situated in a plane is considered a rational point if both its coordinates are rational numbers. Similarly, a line is classified as a rational line when its equation can be expressed using rational numbers. Elliptic curves have an intriguing combination of algebraic structure and geometric properties. When we consider the set of rational points on an elliptic curve, it forms a group, allowing us to perform arithmetic operations with these points. Three points on the elliptic curve "add up to zero" if there is a line that goes through all three of them at once, counting with intersection multiplicities. If a line intersects only two points on the curve, we say that they are additive inverses. And since there's no other point in the plane that works, we create the point O at infinity. This definition gives us some operation, say +, on the points of an elliptic curve. This operation turns out to be associative.

Formally, group addition on an elliptic curve has the following properties:
1. $P + O = O + P = P$ for all $P \in E$
2. $P + (-P) = O$ for all $P \in E$
3. $P + (Q + R) = (P + Q) + R$ for all $P, Q, R \in E$
4. $P + Q = Q + P$ for all $P, Q \in E$

The Mordell–Weil theorem suggests that the group of rational points is always finitely generated (i.e. there always exists a finite set of group generators) for elliptic curves. In other words, there is a finite set of points $P_1, \ldots, P_t \in E(Q)$, $P = n_1 P_1 + n_2 P_2 + \ldots + n_t P_t$ for some $n_1, n_2, \ldots, n_t \in Z$. t would be the rank of this elliptic curve.

Group Addition (Geometric Approach)

Geometrically, group addition on an elliptic curve is done by the tangent-secant method. Following are the steps to do so:

1. Start with a given point L on a cubic curve.

2. Draw the tangent line at point L on the curve.

3. Find the third point of intersection between the tangent line and the curve.

4. If two points, L and M, are given on the curve, draw a line passing through both points.

5. Determine the third point of intersection (x, y) between the line and the curve.

6. The reflection of that point with respect to the x axis, i.e. $(x, -y)$, is $L + M$.

Note that this procedure can be applied to most choices of M and N since cubic curves typically intersect lines at exactly three points. Consider this procedure as a way to "add" two points on the curve and obtain a third point. The following calculates the slope and then the equation of the tangent at the initial points, which eventually leads us to the next points:

```
In[•]:=  slope[{x1_, y1_}, {x2_, y2_}] :=
            If[x1 == x2 && y1 == y2, (3 (x1)^2 + (-5))/(2 * y2), (y2 - y1)/(x2 - x1)];
         intercept[{x1_, y1_}, {x2_, y2_}] := y1 - slope[{x1, y1}, {x2, y2}] * x1;
         nextpoint[{x1_, y1_}, {x2_, y2_}] :=
            {slope[{x1, y1}, {x2, y2}]^2 - x1 - x2, -slope[{x1, y1}, {x2, y2}]^3 +
               slope[{x1, y1}, {x2, y2}] * (x1 + x2) - intercept[{x1, y1}, {x2, y2}]}
```

Adding a Point to Itself

The elliptic curve at hand is $y^1 == x^3 - 5x + 8$. We will find the first rational point L on the curve by using FindInstance:

```
In[•]:=  L = {x, y} /. First[ FindInstance[y^2 == x^3 - 5x + 8, {x, y}, Integers]]
```

```
Out[•]=  {1, -2}
```

```
In[•]:=  nextpoint[L, L]  (* this is 2L *)
```

$$Out[•]= \left\{-\frac{7}{4}, \frac{27}{8}\right\}$$

```
In[•]:=  Show[ContourPlot[{y^2 == x^3 - 5x + 8},
            {x, -3, 5}, {y, -7, 7}, Epilog → {PointSize[0.020], Magenta,
               Point[NestList[nextpoint[#, L] &, L, 1]], {Style[Text["L", L, {0, -1.5}], 15, Black],
                  Style[Text["2L", nextpoint[L, L], {0, -1.5}], 15, Black]}}],
            Plot[slope[L, L] * x + intercept[L, L], {x, -3, 3}, PlotStyle → {Green}],
            Graphics[{Dashed, Line[{{-7/4, 27/8}, {-7/4, -27/8}}]}]]]
```

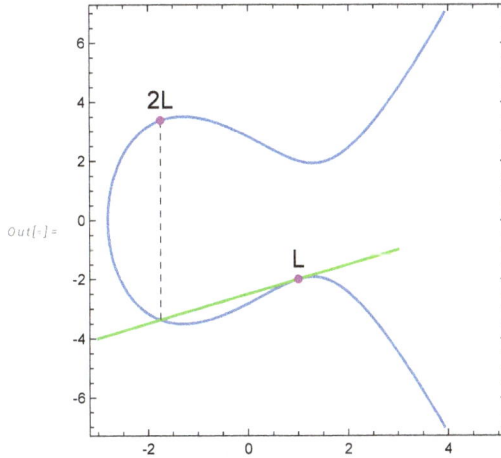

Adding Two Points

```
In[•]:=  M = nextpoint[L, L]
```

$$Out[•]= \left\{-\frac{7}{4}, \frac{27}{8}\right\}$$

In[ ]:= **nextpoint[L, M] (* this is M+L*)**

Out[ ]= $\left\{\dfrac{553}{121}, \dfrac{11\,950}{1331}\right\}$

In[ ]:= **Show[ContourPlot[{y^2 == x^3 −5 x + 8},**
    {x, −3, 5}, {y, −10, 10.5}, Epilog → {PointSize[0.018], Magenta,
      Point[NestList[nextpoint[♯, L] &, L, 2]], {Style[Text["L", L, {0, −1.5}], 15, Black],
       Style[Text["M+L", nextpoint[L, M], {1.2, −0.5}], 15, Black],
       Style[Text["M", nextpoint[L, L], {0, −1.5}], 15, Black]}}],
     Plot[slope[L, M]∗x + intercept[L, M], {x, −10, 5}, PlotStyle → {Green}],
     Graphics[{Dashed, Line[{{$\dfrac{553}{121}$, $\dfrac{11\,950}{1331}$}, {$\dfrac{553}{121}$, $\dfrac{−11\,950}{1331}$}}]}]]]

Out[ ]=

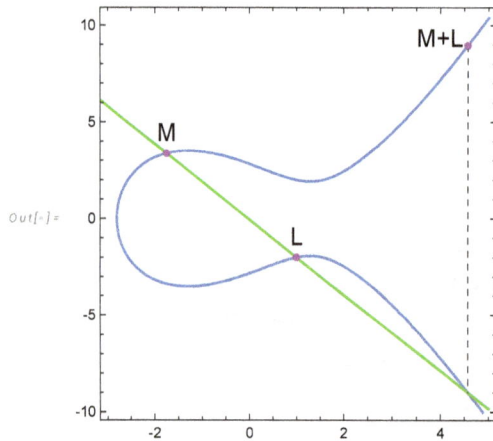

The following is the list of L, 2L, 3L, 4L, …, 8L:

In[ ]:= **NestList[nextpoint[♯, {1, −2}] &, {1, −2}, 8]**

Out[ ]= $\left\{\{1, -2\}, \left\{-\dfrac{7}{4}, \dfrac{27}{8}\right\}, \left\{\dfrac{553}{121}, \dfrac{11\,950}{1331}\right\}, \left\{\dfrac{45\,313}{11\,664}, -\dfrac{8\,655\,103}{1\,259\,712}\right\},\right.$

$\left\{-\dfrac{19\,035\,719}{9\,357\,481}, -\dfrac{89\,393\,659\,342}{28\,624\,534\,379}\right\}, \left\{\dfrac{20\,238\,131\,321}{17\,279\,102\,500}, \dfrac{4\,398\,722\,004\,568\,869}{2\,271\,338\,023\,625\,000}\right\},$

$\left\{\dfrac{492\,427\,847\,046\,961}{935\,706\,047\,761}, -\dfrac{10\,927\,227\,837\,185\,280\,995\,618}{905\,126\,238\,414\,122\,759}\right\},$

$\left\{\dfrac{2\,870\,103\,992\,322\,043\,393}{3\,495\,038\,655\,277\,053\,504}, -\dfrac{13\,780\,060\,468\,451\,643\,045\,994\,142\,977}{6\,533\,982\,622\,887\,348\,588\,544\,276\,992}\right\},$

$\left.\left\{-\dfrac{605\,072\,008\,062\,695\,815\,697\,327}{417\,378\,228\,875\,619\,582\,894\,961}, \dfrac{941\,902\,861\,661\,639\,755\,111\,368\,307\,348\,453\,150}{269\,646\,438\,028\,348\,861\,920\,971\,568\,936\,899\,209}\right\}\right\}$

The following shows the elliptic curve $y^2 == x^3 - 5\,x + 8$ along with 100 rational points L, 2L, 3L, …, 100L, where L = (1, −2):

In[ ]:= **Manipulate[ContourPlot[{y^2 == x^3 − 5 x + 8}, {x, −4, 12}, {y, −20, 20},**
     Epilog → {PointSize[0.015], Magenta, Point[NestList[nextpoint[♯, L] &, L, n]]},
     PerformanceGoal → "Quality"], {n, 1, 100, 1}, SaveDefinitions → True]

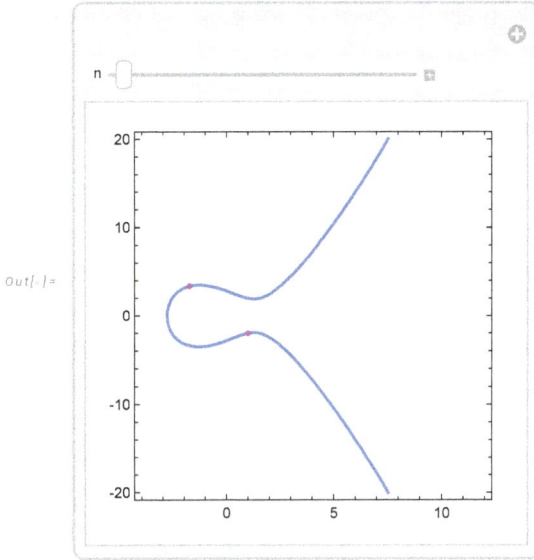

Using Rational Points on an Elliptic Curve to Solve a Diophantine Equation

We will deal with the primary quandary of this project in this section. The question is to find the least positive integers that satisfy the Diophantine equation $\frac{a}{b+c} + \frac{b}{a+c} + \frac{c}{a+b} = n$. We will be dealing with $n=6$ to illustrate the particular solution. To determine the degree of this equation, we multiply by the denominators:

$$n(a+b)(b+c)(c+a) = a(a+b)(c+a) + b(b+c)(a+b) + c(c+a)(b+c) \ [1]$$

Without expanding, we can observe that the highest power of this equation is three.

To place this problem in the right context, we first observe that this is a homogeneous equation. That means if (a, b, c) is a solution, then so is (ka, kb, kc) for some constant value of k. Because multiplying each element by some constant number does not change anything, the constant cancels itself out in each of the fractions:

$$\frac{ka}{kb+kc} + \frac{kb}{ka+kc} + \frac{kc}{ka+kb} = \frac{ka}{k(b+c)} + \frac{kb}{k(a+c)} + \frac{kc}{k(a+b)} = \frac{a}{b+c} + \frac{b}{a+c} + \frac{c}{a+b} = n$$

This shows that, though the equation is three-dimensional, geometrically it can be projected onto a two-dimensional surface. Generally speaking, integer solutions of homogenous equations correspond to rational solutions of the unhomogenized version, which is one dimension lower.

Simplifying equation 1, we get:

$$a^3 + b^3 + c^3 + abc - (n-1)(a+b)(b+c)(c+a) = 0$$

This describes a smooth cubic curve in the projective plane. To bring this into Weierstrass's form for an elliptic curve, we will use the following transformation [4]:

$$x = \frac{-4\,(a+b+2c)\,(n+3)}{(2a+2b-c)+(a+b)\,n} \quad \text{and} \quad y = \frac{4\,(a-b)\,(n+3)\,(2n+5)}{(2a+2b-c)+(a+b)\,n}$$

This leads us to an elliptic curve equation:

$$y^2 = x^3 + (4n^2 + 12n - 3)\,x^2 + 32\,(n+3)\,x$$

It also gives us an inverse transformation:

$$\frac{a}{a+b+c} = \frac{8\,(n+3)-x+y}{2\,(4-x)\,(n+3)}, \quad \frac{b}{a+b+c} = \frac{8\,(n+3)-x-y}{2\,(4-x)\,(n+3)}, \quad \frac{c}{a+b+c} = \frac{-4\,(n+3)-(n+2)\,x}{(4-x)\,(n+3)}, \quad \frac{c}{a+b+c} = \frac{-4\,(n+3)-(n+2)\,x}{(4-x)\,(n+3)}$$

These maps, from a, b and c to x, y and vice versa, show that these two equations are "the same" from the perspective of number theory. Rational solutions of one will give integer solutions of the other. The technical term is birational equivalence, and it's a very fundamental concept in algebraic geometry. There could be some exceptional points that don't genuinely map correctly; those are the cases where $a+b$, $a+c$ or $b+c$ turn out to be 0. This is the standard cost of birational equivalences and shouldn't cause any concern. The specific elliptic curves we are going to deal with have rank 1, which means they have infinitely many rational points, but they are all generated from a single one.

Turning a Diophantine Equation to an Elliptic Curve (Algebraic Expressions)

To get rid of the denominators and express the Diophantine equation in its cubic form, we will multiply out the denominators on both sides and cancel out common terms:

```
In[·]:= Step1 = Assuming[(a + b) (b + c) (a + c) ≠ 0,
            MultiplySides[ a/(b+c) + b/(a+c) + c/(a+b) - n == 0, (a + b) (b + c) (a + c)]] // Together
```

$$\text{Out[·]= } a^3 + a^2\,b + a\,b^2 + b^3 + a^2\,c + 3\,a\,b\,c + b^2\,c + a\,c^2 + b\,c^2 + c^3 - a^2\,b\,n - a\,b^2\,n - a^2\,c\,n - 2\,a\,b\,c\,n - b^2\,c\,n - a\,c^2\,n - b\,c^2\,n == 0$$

Replace a, b, c by their transformations in terms of x and y:

```
In[·]:= Step2 = Step1 /. {a → (8 (n+3)-x+y)/(2 (4-x) (n+3)), b → (8 (n+3)-x-y)/(2 (4-x) (n+3)), c → (-4 (n+3)-(n+2) x)/((4-x) (n+3))} // Factor
```

$$\text{Out[·]= } \frac{(5+2n)\,(96\,x+32\,n\,x-3\,x^2+12\,n\,x^2+4\,n^2\,x^2+x^3-y^2)}{(3+n)^2\,(-4+x)^3} == 0$$

Multiply through by the denominators and factor terms:

$In[\cdot]:=$ **Step3 = Assuming[$\dfrac{(3+n)^2\,(-4+x)^3}{5+2\,n}$ ≠ 0, MultiplySides[Step2, $\dfrac{(3+n)^2\,(-4+x)^3}{5+2\,n}$]] // Factor**

$Out[\cdot]=$ $96\,x + 32\,n\,x - 3\,x^2 + 12\,n\,x^2 + 4\,n^2\,x^2 + x^3 - y^2 == 0$

Obtain the elliptic curve for general n:

$In[\cdot]:=$ **DiophantineToElliptic[n_] = Collect[Reverse@AddSides[Step3, y^2], x]**

$Out[\cdot]=$ $y^2 == (96 + 32\,n)\,x + (-3 + 12\,n + 4\,n^2)\,x^2 + x^3$

Elliptic Curve for Diophantine Equation with $n=6$

The corresponding elliptic curve equation for $\dfrac{a}{b+c} + \dfrac{b}{a+c} + \dfrac{c}{a+b} = 6$ is as follows:

$In[\cdot]:=$ **curve = DiophantineToElliptic[6]**

$Out[\cdot]=$ $y^2 == 288\,x + 213\,x^2 + x^3$

Determine the parameters A and B:

$In[\cdot]:=$ **A := Coefficient[curve[[2]], x^2]**
    B := Coefficient[curve[[2]], x]

Find a rational point on the elliptic curve:

$In[\cdot]:=$ **P = {x, y} /. First[FindInstance[curve && x ≠ 4 && y ≠ 0, {x, y}, Integers]]**

$Out[\cdot]=$ **{−8, 104}**

As mentioned previously, this elliptic curve is of rank 1 [2]. That is, we can generate all the rational points with P. So we will keep on producing $2P$, $3P$, ... using the tangent-secant method. Later, we will check if each one of those corresponds to a positive integer solution for the original Diophantine equation.

Rational Points on the Elliptic Curve $y^2 = 288x + 213x^2 + x^3$

Using EllipticLog and EllipticExp

The value of EllipticLog at the product point equals the sum of values of EllipticLog at the corresponding factors, so we can use EllipticExp to compute the elliptic curve:

$In[\cdot]:=$ **p_⊕q_ := EllipticExp[EllipticLog[p, {A, B}] + EllipticLog[q, {A, B}], {A, B}]**

Compute $\mathcal{P} \oplus \mathcal{P} = 2\mathcal{P}$:

In[ ]:= **P⊕P // N // Chop // Rationalize**

Out[ ]= $\left\{ \dfrac{196}{169}, \dfrac{54796}{2197} \right\}$

Compute $\mathcal{P} \oplus \mathcal{P} \oplus \mathcal{P} = 3\mathcal{P}$:

In[ ]:= **N[P⊕%, 30] // Chop // Rationalize**

Out[ ]= $\left\{ -\dfrac{19719200}{149769}, -\dfrac{67892477120}{57960603} \right\}$

Computing Rational Points Using the Tangent-Secant Method

The following are generalised versions of the duplication formulas used in earlier sections:

In[ ]:= **Slope[{x1_, y1_}, {x2_, y2_}] :=**
    If[x1 == x2 && y1 == y2, ImplicitD[curve, y, x] /. {x → x1, y → y1}, (y2−y1)/(x2−x1)]

In[ ]:= **Intercept[{x1_, y1_}, {x2_, y2_}] := y1 − Slope[{x1, y1}, {x2, y2}] ∗ x1**

In[ ]:= **p_⊕q_ :=**
    {Slope[p, q]^2 − A − (p+q)[[1]],
      −Slope[p, q]^3 + Slope[p, q] (A + (p+q)[[1]]) − Intercept[p, q]}

Compute $\mathcal{P} \oplus \mathcal{P} = 2\mathcal{P}$:

In[ ]:= **P⊕P**

Out[ ]= $\left\{ \dfrac{196}{169}, \dfrac{54796}{2197} \right\}$

Compute $\mathcal{P}, 2\mathcal{P}, \ldots, 5\mathcal{P}$:

In[ ]:= **pts = NestList[p ⟼ (p⊕P), P, 4]**

Out[ ]= $\left\{ \{-8, 104\}, \left\{ \dfrac{196}{169}, \dfrac{54796}{2197} \right\}, \left\{ -\dfrac{19719200}{149769}, -\dfrac{67892477120}{57960603} \right\}, \right.$

$\left\{ \dfrac{1047335279236}{31714979569}, -\dfrac{2976941675946276004}{5648025566504503} \right\},$

$\left. \left\{ -\dfrac{972057154035240008}{706398025591964161}, -\dfrac{1247913070132885704754930984}{593709796498041032772739841} \right\} \right\}$

Animate the sequence of rational points $n\mathcal{P}$ on $y^2 = x^3 + 213x^2 + 288x$ for $n = 1, 2, \ldots, 50$:

```
In[ ]:=  With[
            {plot1 = ContourPlot[{Evaluate[curve]}, {x, –220, 80}, {y, –1200, 1200}, Axes → True],
             pts = List/@NestList[Function[p, p⊕P], P, 50–1]},
            Manipulate[Show[plot1, ListPlot[Take[pts, n], PlotStyle → AbsolutePointSize[10]]],
             {n, 1, 50, 1}, SaveDefinitions → True]]
```

Out[ ]=

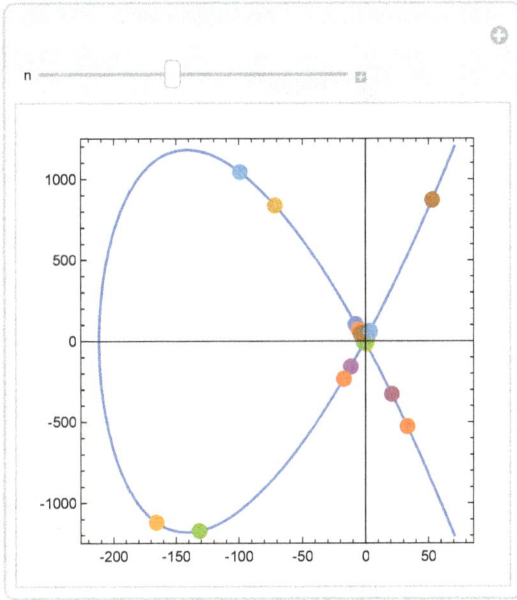

Mapping Rational Solutions of Elliptic Curves Back to the Diophantine Equations

To get back to the original Diophantine equation, we will use the following inverse transformation:

$$\frac{a}{a+b+c} = \frac{8\,(n+3)-x+y}{2\,(4-x)\,(n+3)}, \quad \frac{b}{a+b+c} = \frac{8\,(n+3)-x-y}{2\,(4-x)\,(n+3)}, \quad \frac{c}{a+b+c} = \frac{-4\,(n+3)-(n+2)\,x}{(4-x)\,(n+3)}$$

It is apparent that the inverse transformation formulas described previously only guarantee rational solutions, not integers. So we must multiply them by the least common multiple (LCM) of the denominators of the three fractions to get integer values:

```
In[ ]:=  ellipticToDiophantine[n_, {x_, y_}] :=
            {  8(n+3)–x+y      8(n+3)–x–y      –4(n+3)–(n+2)*x
              ─────────────── , ─────────────── , ─────────────────── }
               2(4–x)*(n+3)      2(4–x)*(n+3)        (4–x)*(n+3)

In[ ]:=  sol = ellipticToDiophantine[6]@NestWhile[p ↦ (p⊕P), P,
                  q ↦ !AllTrue[ellipticToDiophantine[6][q], item ↦ item > 0]];
```

Multiply the solution by the LCM of the denominators of the three fractions to obtain integer values:

```
In[•]:= sol = sol (LCM @@ Denominator[sol])
```

```
Out[•]= {20 260 869 859 883 222 379 931 520 298 326 390 700 152 988 332 214 525 711 323 500 132 179 ⋱.
          943 287 700 005 601 210 288 797 153 868 533 207 131 302 477 269 470 450 828 233 936 557,
          1 218 343 242 702 905 855 792 264 237 868 803 223 073 090 298 310 121 297 526 752 830 558 ⋱.
          323 845 503 910 071 851 999 217 959 704 024 280 699 759 290 559 009 162 035 102 974 023,
          2 250 324 022 012 683 866 886 426 461 942 494 811 141 200 084 921 223 218 461 967 377 588 ⋱.
          564 477 616 220 767 789 632 257 358 521 952 443 049 813 799 712 386 367 623 925 971 447}
```

Check that the solution works:

$$\text{In[•]:= } \frac{a}{b+c} + \frac{b}{a+c} + \frac{c}{a+b} == 6 \text{ /. Thread[\{a, b, c\} → sol]}$$

```
Out[•]= True
```

Positive Solutions of the Diophantine Equation $\frac{a}{b+c} + \frac{b}{a+c} + \frac{c}{a+b} = 6$

In the previous section, we looked at the smallest positive integer solution, which appears at $11P$. Here, we will hunt for all the positive solutions that the rational points yield until $700P$ (where P is $(-8, 104)$). Since this requires calculating the tangent and secant over and over again, it takes a while to generate seven hundred points. This is the code one can use to generate the data:

```
In[•]:= (* Rationals6= NestList[nextRational[#,P]&,P,700];
        Dioph6 = ellipticToDiophantine[6,#]&/@Rationals6;
        PositiveSolutions = Position[AllTrue[#,Positive]&/@ Dioph6, True]; *)
```

If you wish to use the already-generated data, please use the following code. Since it increases the file size a lot, I have only included the visualisation plots of this data:

```
In[•]:= (*Rationals6 =CloudImport[CloudObject["https://wolfr.am/WSS2023–KulkarniData1 "]];*)
        PositiveSolutions =
            CloudImport[CloudObject["https://wolfr.am/WSS2023–KulkarniData2 "]];
```

Following is the plot showing at which multiple of P we get all 26 positive solutions:

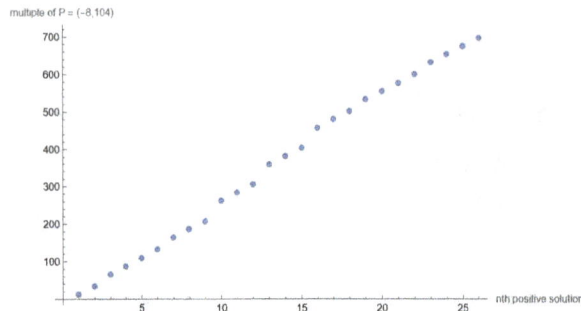

The following plots are of rational points that lead to the positive solutions. It is fascinating to see that all of them are on the first bulblike component of the elliptic curve:

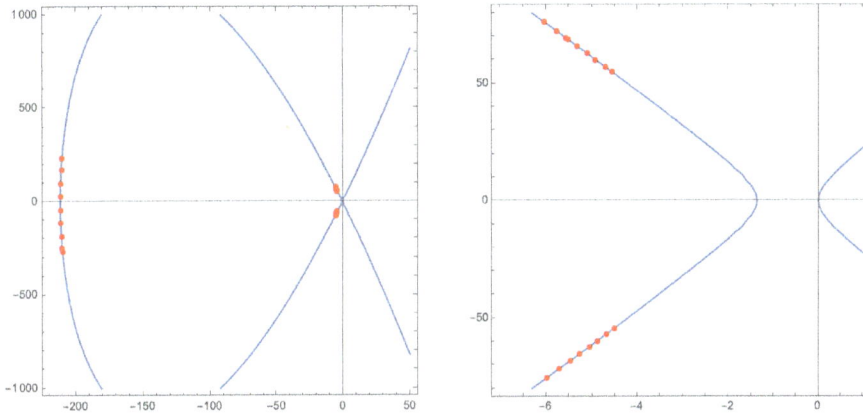

If we take the differences of the multiples (that is, after how many steps will the next positive solution occur?), it's interesting to observe that most times it's 22:

```
In[·]:= Differences[Flatten[PositiveSolutions]]

Out[·]= {22, 32, 22, 22, 22, 32, 22, 22, 54, 22, 22, 54, 22, 22, 54, 22, 22, 32, 22, 22, 22, 32, 22, 22, 22}
```

Solving for Other Values of n

The functions defined in this notebook do not work for all n. They do work for $n = 4$ and $n = 14$. Solutions are given here.

$$\frac{a}{b+c} + \frac{b}{a+c} + \frac{c}{a+b} = 4$$

The minimal solution is:

```
In[·]:= n4 = 4; curve4 = y^2 == x^3 + (4*n4^2 + 12*n4 - 3)*x^2 + 32*(n4+3)*x;
        P4 = {x, y} /. First[FindInstance[curve4 && x ≠ 4 && y ≠ 0, {x, y}, Integers]];
        Slope4[{x1_, y1_}, {x2_, y2_}] :=
          If[x1 == x2 && y1 == y2, ImplicitD[curve4, y, x] /. {x → x1, y → y1}, (y2-y1)/(x2-x1)];
        Intercept4[{x1_, y1_}, {x2_, y2_}] := y1 - Slope4[{x1, y1}, {x2, y2}]*x1;
        nextRational4[{x1_, y1_}, {x2_, y2_}] :=
          {Slope4[{x1, y1}, {x2, y2}]^2 - CoefficientList[curve4[[2]], x][[3]] - x1 - x2,
            -Slope4[{x1, y1}, {x2, y2}]^3 + Slope4[{x1, y1}, {x2, y2}]*
              (CoefficientList[curve4[[2]], x][[3]] + x1 + x2) - Intercept4[{x1, y1}, {x2, y2}]};
        ellipticToDiophantine[n_, {x_, y_}] :=
          {(8 (n+3) - x + y)/(2 (4-x)*(n+3)), (8 (n+3) - x - y)/(2 (4-x)*(n+3)), (-4 (n+3) - (n+2)*x)/((4-x)*(n+3))};
        sol4 = ellipticToDiophantine[n4, NestWhile[nextRational4[#, P4] &, P4,
                  !AllTrue[ellipticToDiophantine[n4, #], Function[item, item > 0]] &]];
        MinSol4 = sol4 *(LCM @@ Denominator[sol4])
```

Out[ ]= {154 476 802 108 746 166 441 951 315 019 919 837 485 664 325 669 565 431 700 026 634 898 253 ⸫
202 035 277 999,

36 875 131 794 129 999 827 197 811 565 225 474 825 492 979 968 971 970 996 283 137 471 637 ⸫
224 634 055 579,

4 373 612 677 928 697 257 861 252 602 371 390 152 816 537 558 161 613 618 621 437 993 378 ⸫
423 467 772 036}

The maximum number of digits in the minimal solution is:

In[ ]:= **Length[IntegerDigits[Max[MinSol4]]]**

Out[ ]= 81

$$\frac{a}{b+c}+\frac{b}{a+c}+\frac{c}{a+b}=14$$

For 14, the maximum number of digits in the minimal solution is:

In[ ]:= **n14 = 14; curve14 = y^2 == x^3+(4*n14^2+12*n14−3)*x^2+32*(n14+3)*x;**
P14 = {x, y} /. First[FindInstance[curve14 && x ≠ 4 && y ≠ 0, {x, y}, Integers]];
Slope14[{x1_, y1_}, {x2_, y2_}] :=
 If[x1 == x2 && y1 == y2, ImplicitD[curve14, y, x] /. {x → x1, y → y1}, (y2−y1)/(x2−x1)];
Intercept14[{x1_, y1_}, {x2_, y2_}] := y1−Slope14[{x1, y1}, {x2, y2}]*x1;
nextRational14[{x1_, y1_}, {x2_, y2_}] :=
 {Slope14[{x1, y1}, {x2, y2}]^2−CoefficientList[curve14[[2]], x][[3]]−x1−x2,
  −Slope14[{x1, y1}, {x2, y2}]^3+Slope14[{x1, y1}, {x2, y2}]*
   (CoefficientList[curve14[[2]], x][[3]]+x1+x2)−Intercept14[{x1, y1}, {x2, y2}]};
ellipticToDiophantine[n_, {x_, y_}] :=
 $$\left\{\frac{8(n+3)-x+y}{2(4-x)*(n+3)}, \frac{8(n+3)-x-y}{2(4-x)*(n+3)}, \frac{-4(n+3)-(n+2)*x}{(4-x)*(n+3)}\right\};$$
sol14 = ellipticToDiophantine[n14, NestWhile[nextRational14[#, P14] &, P14,
 !AllTrue[ellipticToDiophantine[n14, #], Function[item, item > 0]] &]];
MinSol14 = sol14 *(LCM @@ Denominator[sol14]);
Length[IntegerDigits[Max[MinSol14]]]

Out[ ]= 1876

In[ ]:= **MinSol14 ; (*when printed, 1876 digits take a lot of space in the notebook *)**

Concluding Remarks

We find that the corresponding elliptic curve for $\frac{a}{b+c}+\frac{b}{a+c}+\frac{c}{a+b}=6$ is $y^2=288x+213x^2+x^3$. The rational points on this curve lead us to the least positive integers (each of 134 digits) that satisfy the Diophantine equation. If we start with $P=-8104$ and keep on generating $2P$, $3P$, ..., $700P$, it can be seen that there are 26 positive integer solutions. Fascinatingly, they all lie on the bulblike isolated component of the elliptic curve. I would like to investigate the reasons behind that in future. More about using rational points on elliptic curves to find solutions of Diophantine equations can be found in the references. This project is a primary analysis which leaves room for many more extensions.

Currently, the functions that generate rational points on an elliptic curve take the starting rational point from the FindInstance function. As the value of n increases, FindInstance is unable to find any rational starting point. Currently, this project can solve the Diophantine equation $\frac{a}{b+c}+\frac{b}{a+c}+\frac{c}{a+b}=n$ for $n=4$, $n=6$ and $n=14$. I plan to generalise this function as well as study other kinds of Diophantine equations which use rational points on the elliptic curve to find a minimal solution; the rank of elliptic curves used in this project is one. It would be interesting to look at ways to calculate the rank of a curve and further look at curves of higher ranks.

Acknowledgments

I would like to thank my project mentors John McNally and Daniel Robinson for their support and discussions on practical implementations of many ideas. They helped make the Wolfram Summer School experience rewarding for me. I would also like to thank Paul Abbott and Sotiris Michos for their insights on my project, as well as Mads Bahrami, Emily Carter and Danielle Rommel-Mayer for making the admission process and virtual school experience smooth. And I would like to thank Stephan Wolfram and Jonathan Gorard for the project idea.

References

1. J. H. Silverman and J. T. Tate (2015), *Rational Points on Elliptic Curves.* Springer.

2. A. Bremner and A. Macleod (2014), "An Unusual Cubic Representation Problem," *Annales Mathematicae et Informaticae.* publikacio.uni-eszterhazy.hu/2858/1/AMI_43_from29to41.pdf.

3. T. Andreescu, et al. (2010), *An Introduction to Diophantine Equations*, Birkhäuser.

4. D. Yao (n.d.), "Solving a Diphantine Equation with Elliptic Curves," n.p. simonrs.com/eulercircle/crypto2019/darren-ec.pdf.

5. A. Bremner and R. K. Guy (1997), "Two More Representation Problems," *Proceedings of the Edinburgh Mathematical Society* 40: 1–17. www.cambridge.org/core/services/aop-cambridge -core/content/view/7EA1CC1326A2B0701A55E081613332C1/S0013091500023397a.pdf/two -more-representation-problems.pdf.

Cite This Notebook

"From Elliptic Curves to Diophantine Equations: A Journey through Rational Points"
by Aditi Kulkarni
Wolfram Community, STAFF PICKS, July 13, 2023
community.wolfram.com/groups/-/m/t/2959701

HatGame: A Journey through the Hat Tile Configuration Space

JOHANNES MARTIN

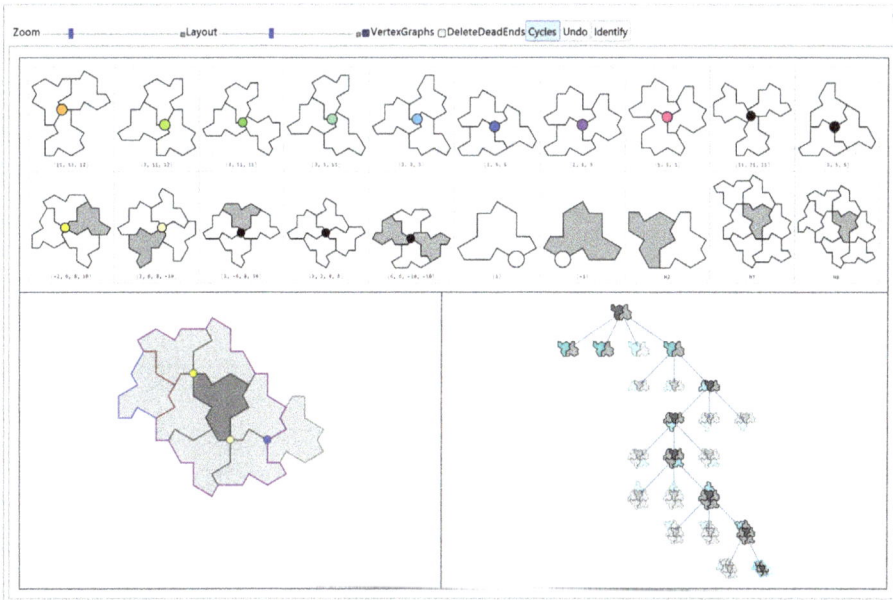

A central objective of this project is the creation of an interactive (one-player) game to explore properties of the hat tile. One can combine jigsaw-like hat tiles into larger and larger clusters. The game contains a few tools that help to analyze the structure of such clusters. First explorations with the HatGame were helpful to cut down the number of possible initial configurations and allowed for extension of "the ten-vertex theorem" of [1]. The hat tile is the first solution that has been found for the so-called "einstein problem," which was the search for one single shape (tile) that can fill the entire plane with only aperiodic patterns (tiling). A tiling is valid and aperiodic when it has the following properties:

1) It covers the entire plane without gaps or holes.

2) There is no translation that maps the pattern onto itself, i.e. there are no translational symmetries.

Remark: a tile is an "Einstein" (an aperiodic mono-tile) when ALL its valid tilings are aperiodic.

Hands-on: The HatGame

In[ ]:=   **HatGame (*!!!Start the game with this function call!!!*)**

Before you dive into this Wolfram Community post, we recommend that you just try to create your own tiling. You do not need to purchase or print your own set of tiles. Just download this notebook and run the function HatGame.

Click any of the patterns of tiles below the control panel as a possible starting configuration. When you click with the mouse at any vertex on the boundary of the initial pattern, the game will try to place another tile at the selected location. In general, there will be more than one possibility to place the tile. These choices are managed inside the tree in the panel of the right-hand side. You can click any of the nodes to switch between choices.

A Tiny Overview

Historical Hat Hunters

To really appreciate the "glory" of the hat tile, we have to turn back time a few decades. In 1961, Hao Wang investigated *formal systems* that could be represented by square tiles with colors:

In[ ]:=   **Wang =** With[...]

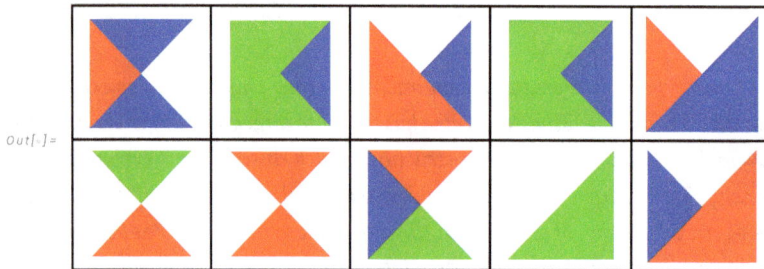

Out[ ]=

The rules of the systems could be restated by the question of whether these squares can be laid out in the plane obeying the rules that:

1. Neighboring squares have to match with their adjacent colors.
2. The resulting pattern of colors should be periodic.

Wang stated the following conjecture: If a finite set of Wang tiles can tile the plane, then there exists a periodic tiling. Due to the finite amount of tiles and periodicity of the tiling, it follows immediately that any periodic tessellation could in principle be confirmed by an algorithm, since it would only require a finite amount of steps.

Five years later, his student Robert Berger showed that any *Turing machine* can be mapped into a set of Wang tiles. Consequently, it follows that any periodic tiling of the entire plane translates into a Turing machine that never stops, i.e. runs forever. Since the *halting problem* is undecidable, there cannot exist an algorithm that decides whether a set of Wang tiles covers the entire plane periodically. Since there is a finite set of Wang tiles that fills the plane, there is only one way to make the problem undecidable. There must also exist aperiodic tilings!

Initially, Berger found a set of 20,426 Wang tiles that covered the plane only in an aperiodical way. Over the course of years, this number was steadily decreased. In 2015, Emmanuel Jeandel and Michael Rao found a set with only 11 tiles and four colors. They also showed that below these numbers, aperiodicity cannot be enforced.

Instead of using different colors, the edges of the squares can be deformed uniquely. Now the color-replacing deformations have to match, as shown in the figure:

In[ ]:= **Graphics[{** ... ✦ **, Polygon[...] ✦ , Darker[...] ✦ ,**
        TranslationTransform[{0.05, 0}] @ Polygon[...] ✦ }, ... → ... ✦]

Out[ ]=

Now the problem of smaller and smaller sets of Wang tiles translates into fewer and fewer different, possibly more complicated shapes of tiles. The quest for the "einstein" had begun. By the way, this phrase has no relation to *the* Albert Einstein—it is just a play on words. The words "*ein Stein*" translate from German into "one piece" or "one stone," and obviously they are used synonymously for the one possible tile that can be used to fill the plane.

Aperiodic tilings made out of only two different shapes were found in 1975 by Roger Penrose. They marked a long-lasting milestone in the quest for the "aperiodic mono-tile." You might expect that the shape of these tiles needs to be bewilderingly peculiar. Well, there exists a Wolfram Demonstrations Project on such Penrose tilings. If you have not seen them before, have a look there. Perhaps you will be surprised by the simple nature of these tiles and how they marvelously fit together, creating a subtle glimpse of non-periodicity: wolfr.am/PenroseTiles

Finally, almost 50 years later—and only a few months ago—this quest has finally come to an end. Let me present to you the first discovered aperiodic mono-tile [4]:

In[ ]:= **Labeled[show[Hat ✦ , Axes → True], Style[...] ✦]**

Out[ ]=

The Hat Tile

The hat tile was discovered by the hobby mathematician David Smith in 2022. Its property of being an aperiodic mono-tile was proven only this year (2023) by a team of mathematicians [4].

You Can Leave Your Hat On

Now that we have given a short story of how the hat tile came into existence and why it is worth thinking about, let's sketch how the HatGame can help us to learn more about this special creature and the aperiodic tilings that it generates.

In the section "Hello Hat":

- We will take a closer look at the shape of the hat tile and break it into eight smaller pieces that are easier to cope with.

- We will introduce a short-hand notation that lays out coordinates in the space of configurations of the hat tile.

- We will give a brief summary about the algorithms that power the game.

In the section "Lessons Learned":

- We will explain how additional constraints and symmetries can be imposed on the space of possible configurations.

- We will also present a few methods that we applied on our journey through the hat tiles' universe.

Hello Hat

The Hidden Shape

At first sight, the shape of the hat tile might seem dizzyingly bizarre and a bit arbitrary. However, it is less random than you would expect from just looking at it. In fact, it is made of eight congruent kites. These kites are uncovered, while the hat tile is covered with two sets of grid lines. The first set is a hexagonal grid that looks like a honeycomb. The second grid is obtained when all midpoints of the hexagons are connected by grid lines. These grid lines break down the hat tile into eight kites of like shape:

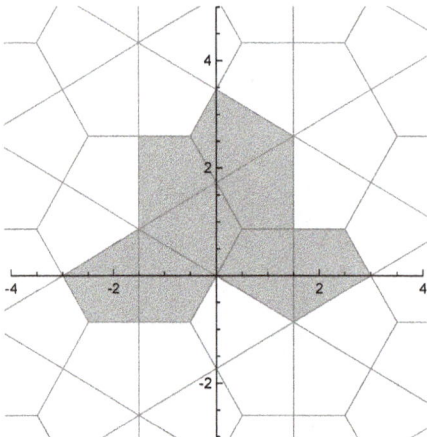

For the sake of completeness, we also show a coordinate representation of the hat tile. You probably never want to work with these coordinates explicitly, and they are not really needed. The hat tiles in this notebook are always generated by a function. The input for this function is a set of numbers that stores all pieces of information. The arguments capture where the hat tile is located, how it is rotated and how it is reflected. The last number of the "code" marks the pivot point among the set of vertices. This code is explained in the next section:

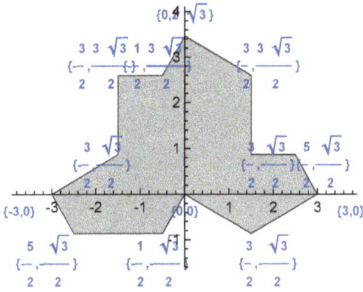

Just in case someone wants to know the coordinates of the 13 points, here they are.

A Convenient Code

Since we want to handle a large number of hat tiles conveniently inside Wolfram Language, we somehow have to store the information of each tile as concisely as possible. There are three different pieces of information needed: the state of rotation, the state of reflection and the position of one of the hat's vertices. These three pieces are collected inside a list of the form {pos, rotation, reflection, vertex}. The hat tile can only be rotated in steps of 30°, so rotations can be represented by integers from 0 to 11. There are only two states of reflection, True or False. And finally, the vertices are labeled by integers 0 to 12. Let's have a look at tiles with the codes:

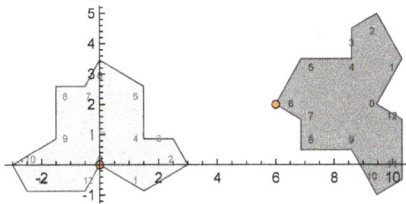

The tile on the left-hand side is in its "normal" form. The vertex 0 is placed at the origin, and there is no rotation and no reflection. The second tile is rotated by 3×30° and the vertex with index 6 is selected to be the pivot point, located at the position provided by the first entry of the code.

How Hats Fit Together

If a person wants to build a large jigsaw out of hat tiles, they can just bring the tiles close together and rotate them until they fit together. This all takes place under the supervision of the person's neural network, formerly known as the brain. The feedback is provided by an arcane sensory system heavily reliant on the presence of electromagnetic radiation in a ridiculously small window of frequencies. In fact, during one of the dinner meetings, we discussed the idea to simulate this process. One would have to build and train a neural network, which could eventually learn—similar to a game of chess—which local moves would pay off in the end and create a space-filling tiling. But due to time constraints, we fell back to a more simple-minded and brute-force implementation that just computes the possible overlap between a tile and an existing cluster of tiles. We investigated a few methods, such as making use of triangulations or index theorems, but we always ran into trouble with vertices that were located on edges.

Last but not least, we implemented a sophisticated polygon fragmentation algorithm that is partially described in [2]. In the tiling game, the user selects a vertex where they want to put the next tile. In a simple preselection process, all configurations of tiles are collected that match with the boundary of the cluster to cover all 360 degrees. These configurations are checked with the polygon fragmentation algorithm; the results are shown in the following table:

```
In[•]:= cluster = Polygon[                Number of points:  28    ];
                                          Embedding dimension:  2

       (*index=19;
         (* index of the point of the cluster, where the hat tile is going to be added*)
         locations =Sort[Mod[findOpenLocs[cluster,index],12]];
         cases=Flatten[Outer[{cluster[[1,index]],#1,#2,#3}&,Range[0,11],
              {True,False}, Keys[Length[locations]/.openLocationsAngleMap],1],2];
         reduced=
           cases[[Flatten[Position[HatAnglesCovered@@@(Rest[#]&/@cases),locations]]]];
         candidates=Tiles@@@reduced;
         result=PlanarPolygonFragmentation[cluster,#]&/@candidates;//AbsoluteTiming*)

Out[•]= {0.474713, Null}
```

This is a part of the function GrowTiling, which is invoked whenever the user wants to add another hat. The heart of the game is calling the PlanarPolygonFragmentation function, developed and kindly provided by Brad Klee. It reliably detects overlapping hat tiles and selects admissible rotations and reflections, even when vertices of the new tile and cluster only meet at the edges. It involves the solution of many intersection problems and is therefore rather expensive. To compensate for this inconvenience, a preselection is performed. First, the range of angles that are not covered by the cluster at the selected vertex point (19 in our example) is determined. The open angles are stored in the list location ({6, 7, 8, 9}):

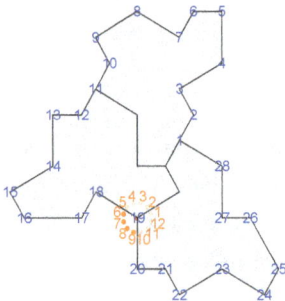

Let us see how the algorithm works for this particular example. The hat tile covers interior angles of 120° at five different vertices {1, 3, 5, 11, 12}. Combined with 12 possible rotations and two reflections, this leads to a total of 120 possible states that are collected in the list "cases." But only 10 of them match with the open angles {6, 7, 8, 9}. From these matches stored in the list "reduce," finally the polygons are constructed. This list of polygons, named "candidates," is fed one by one into the PlanarPolygonFragmentation function. From the 10 results, a selection is shown in the following table:

This table shows a selection of cases that are investigated at a particular point (highlighted in orange) by the action of the PlanarPolygonFragmentation function. We are looking for the situation where there is no intersection. These cases are colored in green. It is also a good place for a warning: When you want to explore tilings exhaustively with the HatGame, you must not select the vertices {11, 13, 9, 8, 5, 2, 28, ...}. At these vertices, two more tiles can be placed, and this possibility is not taken into account by this version of the HatGame. Make sure that you only select vertices where the cluster covers 120° more than 180°.

Lessons Learned (while Playing the HatGame)

Local No-Gos

Although hat tiles may perfectly fit together locally, there is no guarantee that one is able to continue the tiling in such a way that the full plane will be covered eventually. Therefore, it is important to find additional local constraints that have to be fulfilled in every valid tessellation. Let's just focus on vertex combinations with three hat tiles. The color codes for the 10 possible configurations have been taken from [1]. The numbers in the labels are the collection of vertex indices. Each of them corresponds to a different hat:

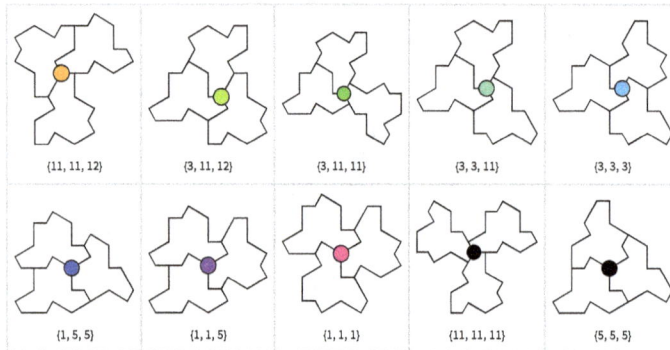

| {11, 11, 12} | {3, 11, 12} | {3, 11, 11} | {3, 3, 11} | {3, 3, 3} |
| {1, 5, 5} | {1, 1, 5} | {1, 1, 1} | {11, 11, 11} | {5, 5, 5} |

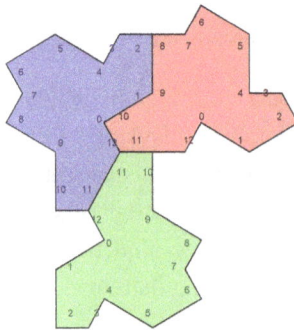

11–11–12

At the common point of intersection, the red and green tiles meet with their vertex with the label 11, whereas the blue tile is located with the vertex 12 at the intersection point.

If you investigate which of these combinations actually occur in valid tilings—as they can, for instance, be generated with the resource functions in [3]—only the first eight of them can be found. With the tiling game, it is now possible to prove that the last two vertex combinations (11-11-11) and (5-5-5) do not lead to tilings that can be extended to infinity.

Let's start with the (11-11-11) vertex combination. The concave regions are uniquely filled with the reflected version of the hat tile:

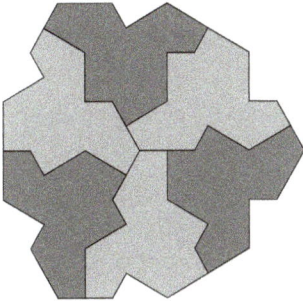

There are a few degrees of freedom for the next tile. However, after a few more tiles, all of the first possible choices will lead to a dead end, meaning that the tiling cannot be completed to cover the entire Euclidean plane. The following graph captures the exhaustive search of all possible tilings:

In[ ]:= **data**

In[ ]:= **(*Rasterize[DecisionTree[data,{{0}},8,17.5,**
      ImageSize→600,"VertexGraphs"→True,ImagePadding→20]]*)

Out[ ]=

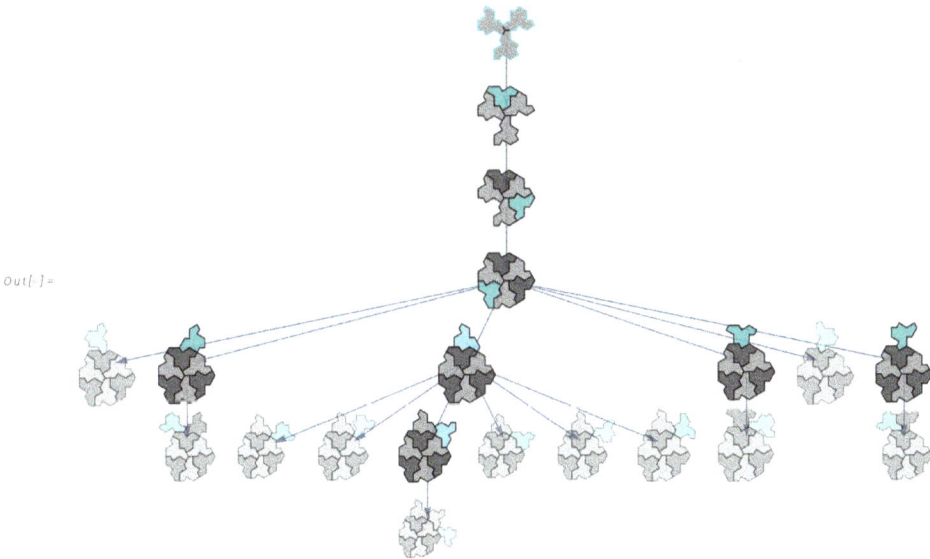

This graph illustrates that the locally valid combination (11-11-11) of initial tiles cannot be completed to a valid tiling that covers the full two-dimensional plane. Up to the fourth level, the continuation of the tiling is unique. The fifth row includes all possible choices that can be made on the top of the tiling at the parent node. In the rest of the graph, it is shown that each of the possible choices will eventually lead to a dead end.

Unfortunately, this process cannot be applied as easily for the (5-5-5) vertex combination. It turns out that this vertex combination can be grown consistently to a very large cluster size. Therefore, some additional simplifications have to be utilized. We can observe that the initial pattern exhibits rotational symmetry. Consequently, any generated nonsymmetric pattern will occur three times, always rotated by 120°:

```
In[·]:=  data2 = ◁|...|▷ ⊛ ;
         (*DecisionTree[data2,{{0}},8,10,ImageSize→600,
            EdgeShapeFunction→({Arrowheads[0.01],Arrow[#1,0.3]}&),
            ImagePadding→50,"VertexGraphs"→True]*)
```

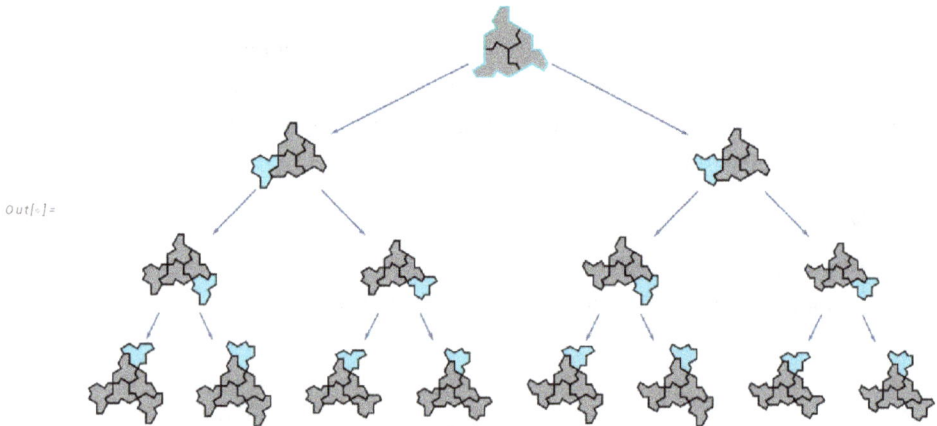

The tree shows the growth of the initial vertex configuration (5-5-5) after two hat tiles have been added.

Let's focus on the patterns in the last row. It's not so easy to spot, but the fourth and fifth patterns are rotationally symmetric. The first, fifth and seventh are related by a rotation of 120° or 240°, and similar relations hold for the second, third and eighth patterns. Don't be distracted by the cyan color of the most recent tile. If you imagine it as gray, the second and third are actually rotationally symmetric. Everything that is observed on the second branch will be observed on the third branch, only rotated by 120°. Without loss of generality, the branches of these related patterns can be combined into one branch. This cuts the amount of bookkeeping to half:

```
In[ ]:= (*data3= <|...|> ;
    DecisionTree[data3,{{0}},8,15,ImageSize→600,ImagePadding→50,
        EdgeShapeFunction→({Arrowheads[0.01`],Arrow[#1,0.2`]}&),
        "VertexGraphs"→True,"highlightLast"→False]*)
```

Reduce the space of configurations by exploiting symmetries. The common shape always displays the configuration of the last branch. To make the symmetry more apparent, the information about the latest hat is made invisible. Out of eight possible configurations, only four of them have to be investigated. The remaining ones are related by rotational symmetry. In what follows, each of the cases will be explored independently, denoted in the order of their appearance in the last row as cases I, II, III and IV.

Let us start with the symmetric cases. The information of every configuration is stored in the global association graphData. Each case can be investigated separately. One only has to reset the {{0}} state of this association with the state that one wants to take a closer look at.

■ Case III

After a few tiles, a dead end is reached for the first of the symmetric cases. The large concave region cannot be filled:

```
In[ ]:= case3 = <|...|> ; PlotTiling[case3, ... ]
```

■ Case IV

Although it is grown much further than the configuration of case III, this symmetric tiling also reaches a dead end at the positions indicated by the arrows, and it therefore cannot give rise to an aperiodic tiling of the plane:

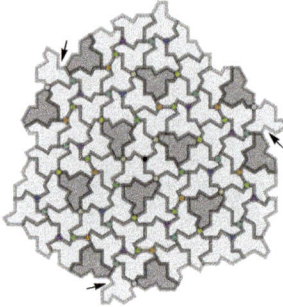

■ Case II

This case quickly runs into trouble, and all possible combinations are readily exhausted. You might argue that it is possible to continue the tiling in a lot of places, and of course you would be right. But the important point is that each configuration contains at least one area where the tiling cannot be continued into. This is all we need, since, after all, we are interested in tiling that can grow everywhere without holes or gaps:

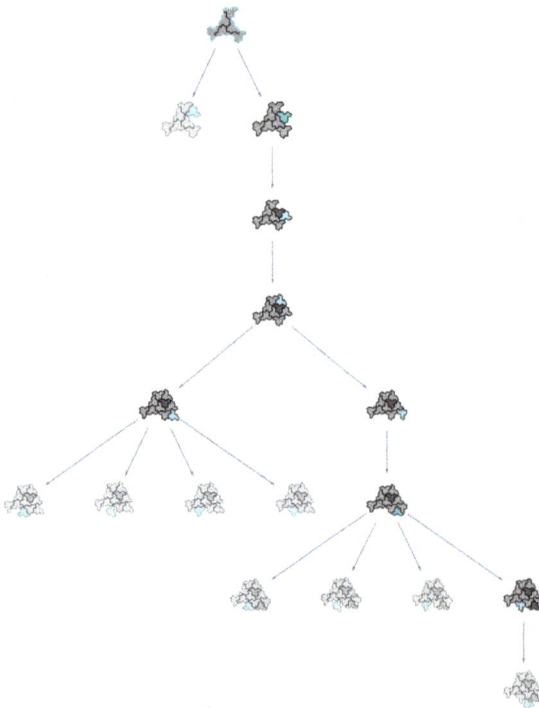

■ Case I

Now the fate of the vertex combination (5-5-5) seems to be almost sealed, and it looks like a trivial exercise to deliver the death blow. However, for a short period of time, we have had a glimpse of hope that there was something exciting to discover. The tiling could be grown consistently and, more importantly, uniquely turned into a rotationally symmetric tiling. The pieces fell so nicely into place that it was really a big disappointment when suddenly the intricately woven tissue was torn apart in two places. Well, *c'est la vie*, and this is especially true for mathematics. You cannot count on something to be persistent or even eternal when you have only watched it for a brief period of time. When you want to talk about infinity, you have to aim for hidden structures. You have to explore patterns and you have to find rules that can be applied infinitely, often as it is presented in the analyses of [4] and [3]:

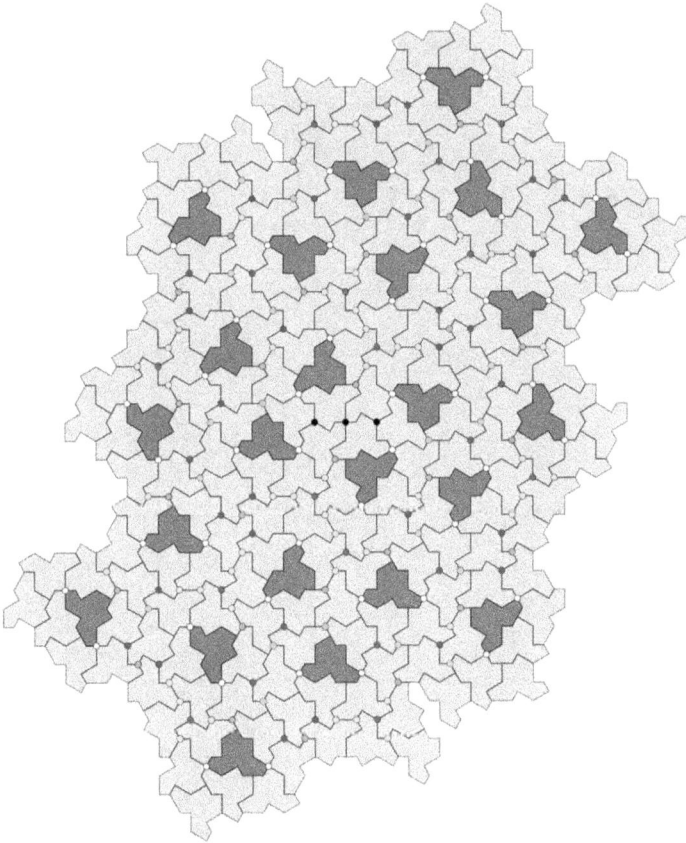

This configuration can be rotated by 180° around the central (2-2-8-8) vertex combination. It is the largest configuration that is unique. For any further tiles, there are at least two choices.

Is the Hat Tiling Unique?

The energy that we spent to disprove the appearance of the (5-5-5) and (11-11-11) vertex configurations was well invested. We can now use these results to reduce the number of possible branches when the tilings are built from valid start configurations. My first choice for this quest is the vertex configuration (3-3-3), with the hope that the symmetry might help again to reduce the number of branches even further. Since everybody is now familiar with the procedure, only a schematic version of the tree graph is plotted:

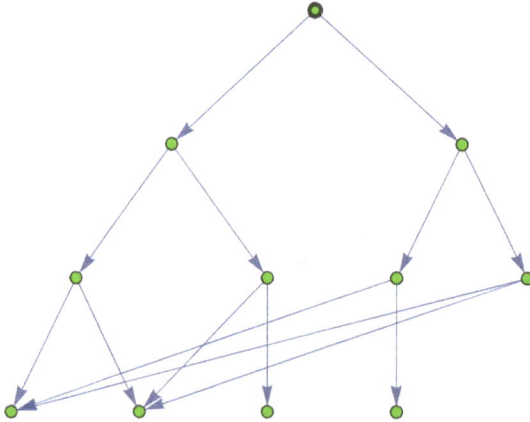

The last states correspond to the following configurations:

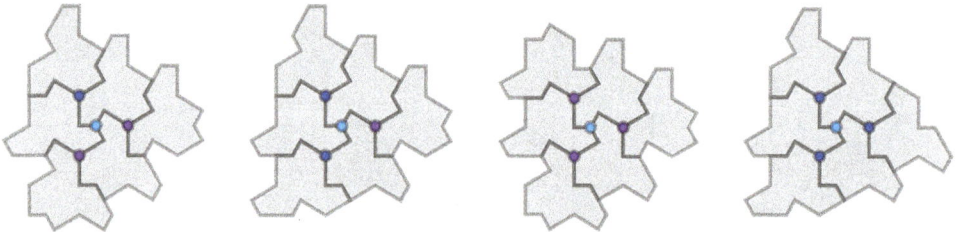

As before, each of the four cases should be investigated separately. The results can be summarized as follows: Unfortunately, even for the symmetric cases, there is usually more than one possibility at each step of the tiling. Many of these branches do not lead to dead ends even after a considerable amount of steps. This makes an exhaustive investigation of the hat tiling at this point impossible. It also indicates that the number of different possible tilings increases with the size of the existing patch. This growing space of possibilities may reflect the aperiodic nature and the fractal-like behavior of the hat tilings. The larger the size of the cluster, the more intricate details have to be generated as you continue.

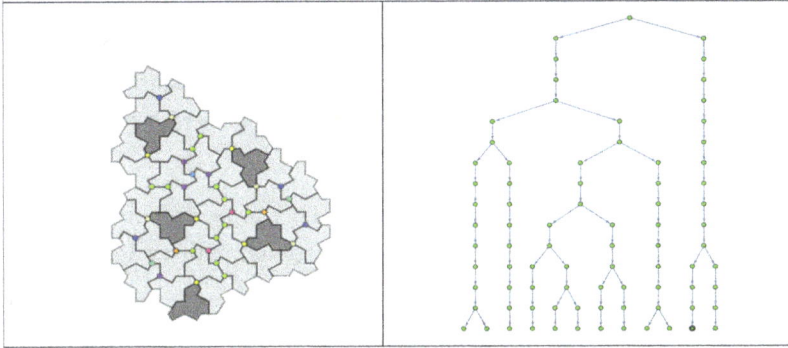

Concluding Remarks

This project was carried out during the three weeks of the Wolfram Summer School 2023. The notes are sketchy and the content always has to be read with a healthy amount of scepticism. It would have been good if there had been more time for a more thorough exploration. But taking into account that the author of this post never had seriously been engaged with any sort of tilings before, the achievements of the three-week boot camp seem quite reasonable. Nevertheless, many statements are a bit vague and they are far from being rigorous results. If you, dear reader, find time to do some explorations on your own, you are invited to do so. Do not hesitate to restate things that we have said as replies to this post.

The graphical user interface that we provide with this project can be viewed as a tool for becoming familiar with the geometric structure of the hat tile and to get a basic feeling of how to start tilings. We can use it to "prove" that certain vertex combinations that are allowed locally will eventually lead to inconsistencies.

For a systematic, bottom-up study of hat tilings, improved tools are required. Due to the growing amount of branches, an automatic search combined with a backtracking algorithm and a fast intersection test seem required. Bowen Ping has developed algorithms that can automatically grow tilings. His results can be found here: wolfr.am/WSS2023-Ping.

Acknowledgments

The foundation for this project was created by Brad Klee, who developed the functionality to investigate intersections of polygons. As our mentor, he was heavily engaged in formulating the objectives and supervising the development and progress of this project. He provided the key to unlock the door to this mathematical wonderland. He created a very pleasant, constructive and interactive atmosphere in our small research group also consisting of Benjamin Peter, Bowen Ping, Russell Martinez and Zsombor Zoltán Méder. Vivid discussions, joint task forces and some relaxing games were both inspiring and rewarding.

This project is closely related to Bowen's project, wolfr.am/WSS2023-Ping. To some extent, we raced side by side trying to uncover the hidden treasures of the hat tile. His help frequently saved me from despair. I also received very helpful support from Mark Greenberg, Bob Nachbar and Eric Parfitt, who patiently tracked down any problems with convoluted lists, graphs and associations.

This project can also be viewed as a first step toward addressing a project proposed by Stephen Wolfram at the 2023 Summer School, "Aperiodic Tilings as Multiway Aggregation."

Last but not least, I want to thank all of the organizers and participants, who for me turned this Summer School into a stimulating, incredibly exciting and unforgettable experience.

References

1. B. Klee (2023), "Hat Combinatorics: The Ten-Vertex Theorem," *Wolfram Community*. community.wolfram.com/groups/-/m/t/2935078.

2. B. Klee (2023), "Checking a Dodecagon Mistake at OEIS," *Wolfram Community*. community.wolfram.com/groups/-/m/t/2932082.

3. B. Klee (2023), HatTrialityTree, Wolfram Language resource function. resources.wolframcloud.com/FunctionRepository/resources/HatTrialityTree.

4. David Smith, et al. (2023), "An Aperiodic Monotile." arXiv preprint (arxiv.org/abs/2303.10798).

5. Joshua E. S. Socolar (2023), "Quasicrystalline Structure of the Smith Monotile Tilings." arXiv preprint (arxiv.org/abs/2305.01174).

Access the Full Code

Scan or visit wolfr.am/WSS2023-Martin.

Cite This Notebook

"HatGame: A Journey through the Hat Tile Configuration Space"
by Johannes Martin
Wolfram Community, STAFF PICKS, July 12, 2023
community.wolfram.com/groups/-/m/t/2958428

Blackbox Optimization: The Mesh Adaptive Direct Search (MADS) Algorithm

MOISES GONZALEZ

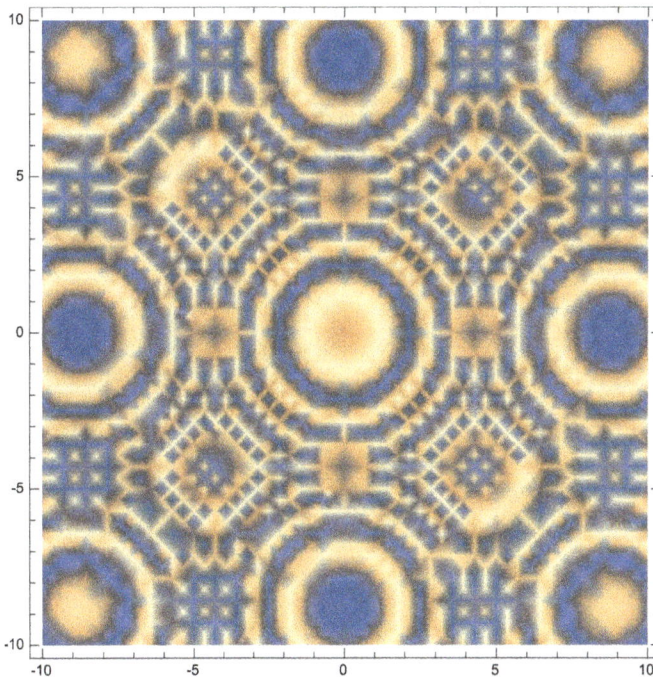

Blackbox optimization is a branch of optimization that deals with derivative-free functions. This means that the function's derivative is unknown because finding it is impossible or impractical. Different methods—like random search, grid search and some heuristics—are used to solve these optimization problems. One recent approach used in this area is the mesh adaptive direct search (MADS) method. This paper aims to explain the creation of a function in Mathematica that runs the MADS algorithm and returns a minimum. This function has been successfully implemented, throwing good results in global minima searches. Nevertheless, using it in functions with many local minima is not suggested because it might get trapped in one of these minima. Also, in the examples, there are functions with problematic derivatives that are successfully solved. The function was named MadsOptimizer, *and it takes two parameters. The first parameter is the objective function, and the second is a list of the variables. Given the positive results of the MADS implementation, the code can be turned into a built-in Wolfram Language function that can be used in different areas. MADS is widely used for parameter fit and other problems involving simulations and experiments. Also, this project suggests an addition to the* MadsOptimizer *function,*

increasing the chances of getting the global minima. This paper grasps some of the theory behind the method, shows some results using different functions and uses a visual representation to show the path taken by the algorithm.

Theory

The main point of this paper is to show the results of the mesh adaptive direct search (MADS) method and why it is a good tool. However, to understand the significance of MADS, it is beneficial to have a basic understanding of its theory. MADS can be seen as a generalization of the generalized pattern search (GPS) optimization method, building upon its principles and introducing new concepts for improved performance. One of the key concepts in the MADS algorithm is the notion of a mesh. In optimization, the mesh refers to the grid or lattice structure that covers the search space. It divides the domain into smaller regions, allowing for a systematic exploration of the parameter space. The mesh helps guide the search by determining the regions to be explored and the precision with which the search is conducted. By adapting the mesh dynamically based on the search progress, MADS can effectively navigate complex and irregular landscapes.

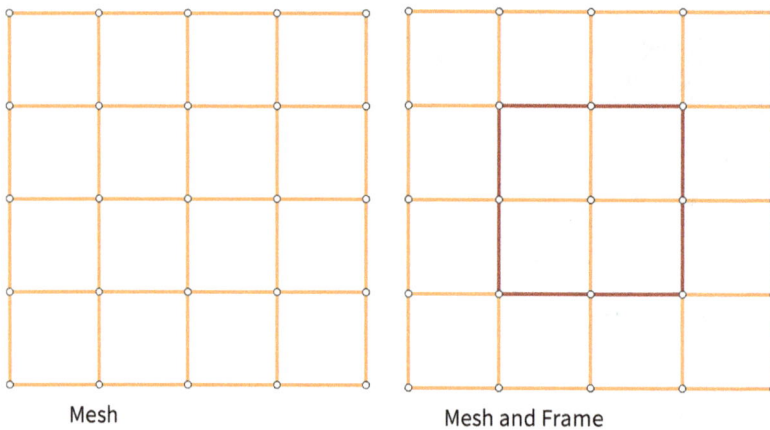

Mesh

Mesh and Frame

Another essential component of the MADS algorithm is the "frame." The frame represents a subset of the mesh, defining the region where the search is currently focused. It acts as a window that constrains the search to a specific part of the parameter space. By adjusting the frame during the optimization process, MADS can adaptively zoom in or out of different areas, allocating computational resources efficiently and directing the search toward promising regions.

The "poll set" is another crucial element of the MADS algorithm. It consists of a collection of trial points carefully selected within the frame. These trial points, which are used to evaluate the objective function, serve as candidates for potential improvement, allowing the algorithm to make informed decisions about which regions of the parameter space to explore further. The poll set plays a pivotal role in balancing exploration and exploitation, enabling MADS to refine the search and converge efficiently toward an optimal solution. Understanding the interplay among these three objects—the mesh, the frame and the poll set—provides insights into the inner workings of the MADS algorithm. While the demonstration of convergence is beyond the scope of this paper, the utilization of these concepts in MADS enables it to achieve improved optimization performance in a variety of real-world problems.

The Code

The main idea behind MADS is to iteratively build and adapt a mesh of trial points in the search space based on the information obtained from function evaluations. The algorithm aims to explore the search space efficiently, avoiding unnecessary function evaluations in regions where the function is unlikely to improve.

The MADS algorithm can be divided into five main steps, so in this section, those steps will be explained through the implementation in Wolfram Language. The first of the steps is the initialization, which initializes the algorithm parameters such as the stopping tolerance, initial frame size parameter and other control parameters like setting the iterator to zero. The function requires the exprn_ and vars_ parameters, which are the function to optimize and its variables, respectively. The user can manipulate the rest of the options if it helps to address a specific problem:

```
In[ ]:=  Options[MadsOptimizer] = {iterations → 10 000,
              meshAdjustment → 0.9, stopTolerance → 10^-60, initialFrameSize → 1};
         MadsOptimizer[exprn_, vars_, OptionsPattern[]] := Module[
            {nDimension = Length[vars],
             kStop = OptionValue[iterations],
             T = OptionValue[meshAdjustment],
             tolerance = OptionValue[stopTolerance],
             DeltaFrame = OptionValue[initialFrameSize],
             kStart = 1,
             deltaMesh}, ...]
```

The second step is simple: update the mesh size parameter, which depends on the frame size parameters. Here it is defined as the minimum value between the frame and the square of the frame, but it can be defined in other ways:

```
In[ ]:=  deltaMeshUpdate = Min[{#, #^2}] &
```

The third step randomly picks a point from the mesh and makes a comparison with the current best point. If the search is successful, the termination step is executed. If the search fails, then the poll step is executed. In this step, the algorithm generates polling points by combining the direction with mesh sizes associated with each direction. Polling points are trial points that will be evaluated to determine if there is an improvement over the current best solution. Moreover, this polling set is an asymptotically dense set of directions, which ensures that the algorithm can approach the optimal solution as the number of iterations or function evaluations tends to infinity. This is necessary to use the Householder matrix and some other operations involving the mesh and frame size parameters. This is a way of picking one direction among the set of directions the frame offers:

```
In[·]:=  HouseholderMatrix[vector_] :=
            IdentityMatrix[Length[vector]] – 2 Transpose[{vector}] . {vector}/(vector . vector)

In[·]:=  RandomPollStep[nDimension_Integer, DeltaFrame_] := Module[
                {vector = Normalize[RandomReal[1, nDimension]],
                 deltaMesh1 = Min[deltaFrame^2, deltaFrame],
                 Hmat,
                 Bset,
                 Dset},
                Hmat = HouseholderMatrix[vector];
                Bset = (Round[(deltaFrame/deltaMesh)*(#/Norm[#, Infinity])] &) /@
                Transpose[Hmat];
                Dset = Transpose@Join[Bset, –Bset];
                Return[Transpose[Dset][[RandomInteger[{1, 2 * nDimension}]]]]
                ]
```

The objective function is evaluated at a randomly selected polling point to obtain the corresponding function value. Then the function value is compared with the current best solution to determine if any improvement is found. If any polling point yields a better solution, the current best solution is accordingly updated. The best solution is typically the point with the lowest function value:

```
In[·]:=  currentPoint = (bestPoint + deltaMesh * RandomPollStep[nDimension, DeltaFrame])
```

Based on the evaluation results of the polling points, adapt the frame sizes associated with each search direction. The frame sizes control the density of trial points along each direction. If a direction leads to improved solutions, the frame size in that direction may be decreased to focus the search. Conversely, if a direction is unproductive, the frame size may be increased to reduce the sampling density:

```
In[·]:=  If[ObjectiveFunction @@ currentPoint < ObjectiveFunction @@ bestPoint,
            (*If True, update the best point and expand the frame size*)
            bestPoint = currentPoint;
            DeltaFrame = DeltaFrame/T,
            (*If False, make smaller the frame*)
            DeltaFrame = DeltaFrame*T]
```

The final step is to check if the termination criteria are met. These criteria can be based on the number of iterations, the established tolerance or a specific target objective value. If the termination criteria are satisfied, stop the algorithm and return the current best solution as the approximate optimal solution. In this implementation, the two termination criteria are iterations and tolerance:

```
In[·]:=  If[DeltaFrame ≥ tolerance,
            kStart++,
            Return[{N[ObjectiveFunction @@ bestPoint], N[bestPoint]}]
         ]
```

In addition to the MADS algorithm, a piece of code was added. This extension is a function called MoxOptimizer and, like MadsOptimizer, has the objective function and respective variables as required parameters. The main idea of this extension is to run the MADS algorithm several times. The objective is to initialize the search with different random vectors and then pick the best of the MADS outcomes, which increases the probability of getting the correct answer. The following example shows the difference between MadsOptimizer and MoxOptimizer:

In[·]:= **MadsOptimizer[−Cos[x] ∗ Cos[y] ∗ e^(−(x − π)^2 − (y − π)^2), {x, y}]**

Out[·]= $\text{MadsOptimizer}\left[-e^{-(-\pi+x)^2-(-\pi+y)^2}\ \text{Cos}[x]\ \text{Cos}[y], \{x, y\}\right]$

In[·]:= **MoxOptimizer[−Cos[x] ∗ Cos[y] ∗ e^(−(x − π)^2 − (y − π)^2), {x, y}]**

Out[·]= $\text{MoxOptimizer}\left[-e^{-(-\pi+x)^2-(-\pi+y)^2}\ \text{Cos}[x]\ \text{Cos}[y], \{x, y\}\right]$

Notice how the MoxOptimizer function is slower than MadsOptimizer; however, MoxOptimizer has more chance to give the correct global minima. This improvement is not rooted in the structure of the second function; instead, the improvement comes because of the change of the initial vectors with each iteration. Also, by augmenting the repetitions option, both running time and the chance of getting the correct answer increase:

In[·]:= **MoxOptimizer[−Cos[x] ∗ Cos[y] ∗ e^(−(x − π)^2 − (y − π)^2), {x, y}, repetitions → 12]**

Out[·]= $\text{MoxOptimizer}\left[-e^{-(-\pi+x)^2-(-\pi+y)^2}\ \text{Cos}[x]\ \text{Cos}[y], \{x, y\}, \text{repetitions} \to 12\right]$

In[·]:= **MoxOptimizer[−Cos[x] ∗ Cos[y] ∗ e^(−(x − π)^2 − (y − π)^2), {x, y}, repetitions → 2]**

Out[·]= $\text{MoxOptimizer}\left[-e^{-(-\pi+x)^2-(-\pi+y)^2}\ \text{Cos}[x]\ \text{Cos}[y], \{x, y\}, \text{repetitions} \to 2\right]$

Testing

After several tests, the MadsOptimizer and MoxOptimizer functions improved over NMinimize in specific cases. One example is the function previously used, which is known as the Easom function and is defined as:

$$f(x, y) = -\cos(x) \cdot \cos(y) \cdot e^{-(x-\pi)^2 - (y-\pi)^2}$$

The next cell contains the result of evaluating the Easom function in NMinimize.

In[ ]:= **NMinimize[–Cos[x] ∗ Cos[y] ∗ 𝑒 ^ (–(x – 𝜋) ^ 2 – (y – 𝜋) ^ 2), {x, y}]**

Out[ ]= {–0.0000811022, {x → 1.305, y → 1.305}}

In contrast, if MadsOptimizer is used, there is a chance of getting the correct answer, which is $f(\pi, \pi) == -1$. In addition to this, if MoxOptimizer is evaluated, the chance of getting the correct answer is high:

In[ ]:= **MadsOptimizer[–Cos[x] ∗ Cos[y] ∗ 𝑒 ^ (–(x – 𝜋) ^ 2 – (y – 𝜋) ^ 2), {x, y}]**

Out[ ]= $\text{MadsOptimizer}\left[-e^{-(-\pi+x)^2-(-\pi+y)^2} \text{Cos[x] Cos[y]}, \{x, y\}\right]$

In[ ]:= **MoxOptimizer[–Cos[x] ∗ Cos[y] ∗ 𝑒 ^ (–(x – 𝜋) ^ 2 – (y – 𝜋) ^ 2), {x, y}]**

Out[ ]= $\text{MoxOptimizer}\left[-e^{-(-\pi+x)^2-(-\pi+y)^2} \text{Cos[x] Cos[y]}, \{x, y\}\right]$

The following graph is a representation of how MadsOptimizer finds the global minima:

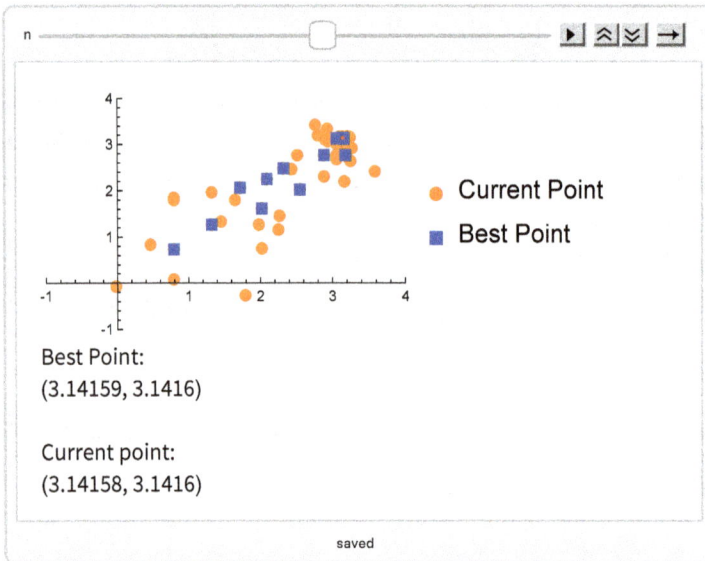

The red dot represents the coordinates of the minima, the blue squares are the best points found and the orange dots are all the points' progressions. Here, the algorithm starts searching points in the mesh and gradually approaches the global minimum. Nevertheless, it is essential to be aware that this behavior may lead to local minima in which the algorithm might get trapped:

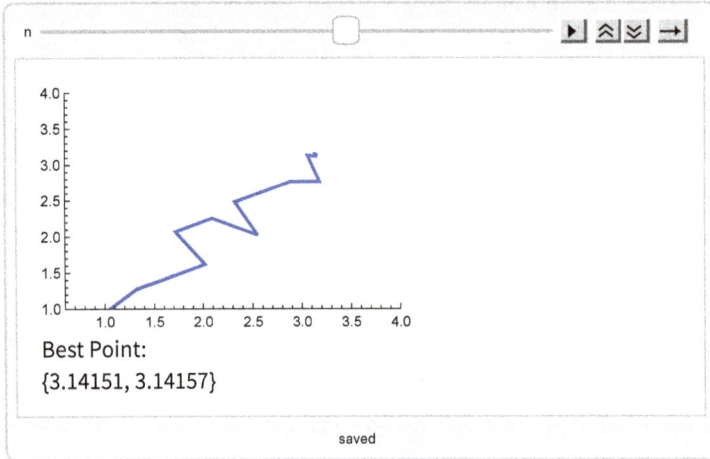

Best Point:
{3.14151, 3.14157}

saved

This other visualization represents the path of the best points connected by straight blue lines. The rest of this notebook is dedicated to show a few examples of some functions evaluated in NMinimize, MadsOptimizer and MoxOptimizer, respectively:

In[ ]:= **(∗ Function = Sin[x^2 + y^2] ∗)**

Print["NMinimize: ", NMinimize[Sin[x ^ 2 + y ^ 2], {x, y}]]
Print["MadsOptimizer: ", MadsOptimizer[Sin[x ^ 2 + y ^ 2], {x, y}]]
Print["MoxOptimizer: ", MoxOptimizer[Sin[x ^ 2 + y ^ 2], {x, y}]]

NMinimize: {0., {x → 0., y → 0.}}

MadsOptimizer: MadsOptimizer$[Sin[x^2 + y^2], \{x, y\}]$

MoxOptimizer: MoxOptimizer$[Sin[x^2 + y^2], \{x, y\}]$

Here, NMinimize does not arrive at the correct answer. On the other hand, MadsOptimizer does arrive at the correct global minima very often, while MoxOptimizer gets the minima most of the time:

In[ ]:= **(∗ Function = Abs[x]∗Sin[y] + Abs[y]∗Sin[x] ∗)**

Print["NMinimize: ", NMinimize[Abs[x] ∗ Sin[y] + Abs[y] ∗ Sin[x], {x, y}]]
Print["MadsOptimizer: ", MadsOptimizer[Abs[x] ∗ Sin[y] + Abs[y] ∗ Sin[x], {x, y}]]
Print["MoxOptimizer: ", MoxOptimizer[Abs[x] ∗ Sin[y] + Abs[y] ∗ Sin[x], {x, y}]]

NMinimize: {−3.63941, {x → −2.02876, y → −2.02876}}

MadsOptimizer: {−3.63941, {−2.02876, −2.02876}}

MoxOptimizer: {−9.79181, {−8.49475, −1.66531}}

The second example shows how MadsOptimizer and NMinimize arrive at the same minima, though MoxOptimizer found a better answer. MoxOptimizer does not always arrive at the same coordinate and might sometimes arrive at the same point as the other two functions. Finally, the third example shows that the three functions arrived at the same minima. One aspect of

NMinimize is that it always has the same outcome, while MadsOptimizer and MoxOptimizer may give different answers for the same function. Nevertheless, MoxOptimizer is usually stable in its outputs because of the repetitions:

In[·]:=   (*   Function = (x^2 + y − 11)^2 + (x + y^2 − 7)^2   *)

Print["NMinimize: ", NMinimize[(x^2+y−11)^2+(x+y^2−7)^2, {x, y}]]
Print["MadsOptimizer: ", MadsOptimizer[(x^2+y−11)^2+(x+y^2−7)^2, {x, y}]]
Print["MoxOptimizer: ", MoxOptimizer[(x^2+y−11)^2+(x+y^2−7)^2, {x, y}]]

NMinimize: {0., {x → 3., y → 2.}}

MadsOptimizer: {4.93038×10^{-32}, {3.58443, −1.84813}}

MoxOptimizer: {0., {3., 2.}}

MadsOptimizer and MoxOptimizer

Access the Full Code

Scan or visit wolfr.am/WSS2023-Gonzalez.

Concluding Remarks

In this project, two functions were developed. The first function is MadsOptimizer, and its code is based on the MADS algorithm. This method is helpful in blackbox optimization, solving derivative-free functions and other functions with problematic derivatives. It can be applied to fields like material science, energy and computer science. One interesting approach is to optimize the hyperparameters in machine learning. Because of the time such an application may take, it was left for future research work. The second function developed depends on the first, a straightforward and logical extension. The MoxOptimizer function will execute the MADS algorithm several times, allowing different starting vectors to help find other solutions. These solutions are stacked in a list in increasing order for the evaluation, and finally, the first element is selected as the solution. Summing up, the algorithm was implemented as a function in Wolfram Language with the name MadsOptimizer. In addition, another function (MoxOptimizer) was created to improve the search of the global minima. The MadsOptimizer function showed some improvement in finding global minima in specific functions when compared with the function NMinimize.

Acknowledgments

I want to thank God and my family, who gave everything. I am also grateful to my mentor Sotiris Michos, who helped me from the project selection until the culmination. I want to thank Christian Pasquel for the guidance since week zero. I thank Robert Nachbar for his availability and great Wolfram Language introduction lectures. I am also thankful to Joshua Pedro, Faizon Zaman, Mark Greenberg, Swastik Banerjee, John McNally, Daniel Sanchez and, in general, all available mentors. Finally, I want to thank Mads Bahrami, Yi Yin, Stephanie Bowyer and Ahmed El Banna for their patience and exceptional guidance through these three weeks. Finally, I am profoundly thankful to Stephen Wolfram for making this fantastic learning experience possible.

References

1. C. Audet and W. Hare (2017), *Derivative-Free and Blackbox Optimization*. Springer.

2. C. Audet and J. E. Dennis (2006), "Mesh Adaptive Direct Search Algorithms for Constrained Optimization," *SIAM Journal on Optimization* 17: 188–217. researchgate.net/publication/220133585_Mesh_Adaptive_Direct_Search_Algorithms_for _Constrained_Optimization.

3. S. Alarie, et al. (2021), "Two Decades of Blackbox Optimization Applications," *EURO Journal on Computational Optimization* 9: 100011. doi: 10.1016/j.ejco.2021.100011.

4. C. Audet and M. Kokkolaras (2016), "Blackbox and Derivative-Free Optimization: Theory, Algorithms and Applications," *Optimization and Engineering* 17: 1–2. doi: 10.1007/s11081-016-9307-4.

Cite This Notebook

"Blackbox Optimization: The Mesh Adaptive Direct Search (MADS) Algorithm"

by Moises Gonzalez

Wolfram Community, STAFF PICKS, July 12, 2023

community.wolfram.com/groups/-/m/t/2958734

www.ingramcontent.com/pod-product-compliance
Lightning Source LLC
Chambersburg PA
CBHW051117200326
41518CB00016B/2527